Synthesis Lectures on Mobile and Pervasive Computing

Editor

Mahadev Satyanarayanan, Carnegie Mellon University Mobile computing and pervasive computing represent major evolutionary steps in distributed systems, a line of research and development that dates back to the mid-1970s. Although many basic principles of distributed system design continue to apply, four key constraints of mobility have forced the development of specialized techniques. These include unpredictable variation in network quality, lowered trust and robustness of mobile elements, limitations on local resources imposed by weight and size constraints, and concern for battery power consumption. Beyond mobile computing lies pervasive (or ubiquitous) computing, whose essence is the creation of environments saturated with computing and communication yet gracefully integrated with human users. A rich collection of topics lies at the intersections of mobile and pervasive computing with many other areas of computer science.

A Practical Guide to Testing Wireless Smartphone Applications Julian Harty

Location Systems: An Introduction to the Technology Behind Location Awareness Anthony LaMarca and Eyal de Lara Replicated Data Management for Mobile Computing Douglas B. Terry

Application Design for Wearable Computing Dan Siewiorek, Asim Smailagic, and Thad Starner Controlling Energy Demand in Mobile Computing Systems Carla Schlatter Ellis RFID Explained

Roy Want

A Practical Guide to Testing Wireless Smartphone Applications
Julian Harty

ISBN: 978-3-031-01351-5 paperback ISBN: 978-3-031-02479-5 ebook

DOI: 10.1007/978-3-031-02479-5

A Publication in the Springer series
LECTURES ON MOBILE AND PERVASIVE COMPUTING

Lecture #6

Series Editor: Mahadev Satyanarayanan, Carnegie Mellon University

Series ISSN

ISSN 1933-9011 print

ISSN 1933-902X electronic

A Practical Guide to Testing Wireless Smartphone Applications

Julian Harty
Google Inc.

SYNTHESIS LECTURES ON MOBILE AND PERVASIVE COMPUTING # 6

ABSTRACT

Testing applications for mobile phones is difficult, time-consuming and hard to do effectively. Many people have limited their testing efforts to hands-on testing of an application on a few physical handsets, and they have to repeat the process every time a new version of the software is ready to test. They may miss many of the permutations of real-world use, and as a consequence their users are left with the unpleasant mess of a failing application on their phone.

Test automation can help to increase the range and scope of testing, while reducing the overhead of manual testing of each version of the software. However automation is not a panacea, particularly for mobile applications, so we need to pick our test automation challenges wisely. This book is intended to help software and test engineers pick appropriately to achieve more; and as a consequence deliver better quality, working software to users.

This Synthesis lecture provides practical advice based on direct experience of using software test automation to help improve the testing of a wide range of mobile phone applications, including the latest AJAX applications. The focus is on applications that rely on a wireless network connection to a remote server, however the principles may apply to other related fields and applications.

We start by explaining terms and some of the key challenges involved in testing smartphone applications. Subsequent chapters describe a type of application e.g. markup, AJAX, Client, followed by a related chapter on how to test each of these applications. Common test automation techniques are covered in a separate chapter, and finally there is a brief chapter on when to test manually.

The book also contains numerous pointers and links to further material to help you to improve your testing using automation appropriately.

KEYWORDS

automation, mobile, test, wireless

Preface

Welcome to the first edition of the *A Practical Guide to Testing Wireless Smartphone Applications.* I hope you find this material useful as a technical introduction to the thorny subject of how to automate tests for mobile wireless applications that run on smartphones. Here you can learn about the basics of various practical techniques, then follow the links and references to learn more about topics of relevance to you. I encourage you to seek other examples and references, and to experiment with test automation, in order to test your applications more effectively.

I wrote this guide as I could not find anything similar when I started in the field of mobile wireless testing. After two years of adding material to my notes the content is still unfinished and incomplete; and I realize I am unlikely to ever "catch up" to cover the encyclopedia of topics, as the field of test automation for wireless smartphone applications continues to change and evolve. So, here is what I have written so far. I hope it will enable others to learn about the subject much more quickly.

I hope to publish updated versions as I learn more about the subject, and I welcome your contributions. Some draft material is also available online at http://sites.google.com/site/mobilewirelesstestautomation/

I currently work at Google. Some of the examples come from issues encountered at work, the rest come from a wide range of sources.

Please be aware that this work continues to evolve based on my experiences: for instance I would like to include information on test automation for the Android platform once I have the relevant experience, which is driven—in part—by my project assignments.

What Is Inside

Chapter 1   We start with an overview of the field of mobile wireless test automation: What is it? What are the challenges? And how will we approach the problem of test automation. I've included a summary of topics currently beyond the scope of this book, or my experience, others are welcome to provide material on these topics.

Chapter 2   Introduces markup languages and provides examples of two common languages: WML and xHTML.

Chapter 3   Introduces some test automation techniques for markup languages. The code is built up from very modest beginnings, and thankfully it remains compact and easy to comprehend after we have enhanced the scripts to do more.

Chapter 4   AJAX applications are starting to be developed specifically for mobile devices, e.g., for the iPhone. This chapter explains the fundamentals of AJAX mobile applications, how they behave, and explains some of the testing challenges.

Chapter 5   The testing strategy for mobile AJAX applications combines testing techniques for desktop AJAX applications with several tricks based on our work for markup applications.

Chapter 6   Introduces client applications and provides a high-level testing strategy for them.

Chapter 7   Includes the testing techniques for client applications.

Chapter 8   Includes common techniques which are broadly applicable across multiple applications.

Conventions

Text is in this font. Code examples have a shaded background, e.g.,

```
import urllib
```

Key lines of code will be in bold, and may be numbered, e.g.,

```
response = request.open("http://www.google.co.uk/m") (1)
```

1. Each numbered line would be described immediately after the example code.

Commands, system responses, and code within the main body of the text, etc., will be formatted in Courier (e.g., man minicom) which you can enter in a suitable terminal command window.

Hyperlinks are live in the electronic edition of this book e.g., http://google.com/m

What You Will Need

All the examples use freely available software. The main programs are: Python, Java, and the Firefox web browser. Recent versions of both tools (e.g., Python 2.4, Java 5, and Firefox 1.5) should be suitable, although I suggest you use the current versions of the programs.

Using Code Examples

All the code samples are available from the author or at http://code.google.com/p/mwta/. You are free to use and modify the code.

Alphabet Soup and the Half-life of Links

On reviewing my writings, I am overwhelmed by the number of acronyms and aware of how quickly the links decay, either by the target disappearing or because they are no longer the most suitable or the most current reference. When you meet an unfamiliar acronym please search the internet for the meaning and derivation. For old and decaying links—make notes on your copy of the book and add updated links and comments to keep your copy current.

Acknowledgments

Thank you to my colleagues at Google who have helped me understand many of the nuances of mobile wireless technologies. Greg Block contributed the material on image stitching. Mike Davis reviewed much of the book for technical content.

Michela Wrong also reviewed the entire book, at short notice and while on holiday. Her diligent notes enabled me to make the content more consistent, clearer, and more readable.

Thank you also to Google who hired me as a novice in the field of mobile wireless testing, they encouraged me to learn the technical nitty-gritty of the domain while I tried to contribute usefully in terms of testing our mobile applications. They also permitted me to release this material for public consumption which helps spread the knowledge and understanding of a fairly specialist field.

Contents

Author Biography

CHAPTER 1

Introduction

1.1 WHAT IS A MOBILE WIRELESS APPLICATION?

I will start at the beginning with a working description of the term I will be using throughout the rest of this work: *mobile wireless application.*

Mobile refers to the intent, the devices are portable, often lightweight devices that move around, often carried by their user. The devices are generally powered with a small battery, which implies a tradeoff between power, functionality, and battery life.

Wireless devices communicate with other devices without physical wires or cables.

Application refers to the software used by the user on the device. The application may be written to run on the mobile device or take advantage of existing software on the device such as a web browser.

The term includes mobile phones and handheld devices that communicate over wireless networks. This book also covers testing of some aspects of the servers that support the mobile wireless applications.

Mobile wireless applications include communications over-the-air (OTA) between the mobile wireless device and servers. Connections are mainly over external mobile phone networks, although WiFi is an option on newer devices.

A brief introduction to mobile network history and terminology is available at http://umtsmon.sourceforge.net/docs/terminology.shtml.

1.2 CLASSIFICATIONS OF MOBILE WIRELESS APPLICATIONS

As I have gained experience with mobile wireless applications, I have come to identify ways to group similar types of these applications using an inform al classification, which has helped me to understand their similarities and differences.

- Client applications. These are split into two groups: native and portable,
- Messaging applications,
- Browser applications. Also split into two groups: markup and AJAX applications.

I will be using these terms throughout the rest of the document. I encourage you to adapt the classification scheme to suit your needs.

1.2.1 Client Applications

Client applications are installed on a mobile device and run on that device.

The application may be written to look and feel like a native application for specific phone models. A native application should behave and look like an integrated part of the installed phone software. Generally, custom compilers and tools are needed to build native software specifically for those phone models.

Portable applications are generally able to run with few changes across a wide range of phone models and manufacturers. The user interface is not as well integrated with any individual phone model and the software may not be able to take advantage of all the features provided by particular phone models.

1.2.2 Messaging Applications

Current messaging applications use SMS messages as the communications medium. Typically the user can use the standard "text messaging" feature provided with the phone.

A single SMS message contains between 70 and 160 characters depending on how the characters are encoded. The protocol has been extended to send longer messages (http://en.wikipedia.org/wiki/SMS).

The servers need to receive and respond to the specifications of SMS messages. The messages are packed and need to be decoded before being used. Virtually every mobile phone includes full support for SMS messaging; and manufacturers provide SMS software libraries if you want to incorporate SMS communication into a custom application.

1.2.3 Browser Applications

Browser applications are server-based applications that can be accessed through a web browser via a URL from a mobile device. There are a variety of web-based markup languages, dictated by the capabilities of the web browsers for different geographic regions, etc. Mobile web browsers are less flexible or capable than desktop web browsers, e.g., they are unlikely to support extensions and media players (such as Flash).

Markup applications are generated and run within the server. The client displays, or renders, the pages generated by the server and provides basic user-interaction. User input is sent by the browser to the server for processing.

Modern mobile web browsers are beginning to have support for AJAX applications—JavaScript that runs within the web browser on the client, and enables developers and designers to create richer applications. The JavaScript often modifies the page content within the browser, and interacts directly with the server.

1.2.4 The Supporting Servers

The servers include varying degrees of customization for mobile wireless applications. For instance, web servers can detect requests from mobile devices and tailor content accordingly. The customization helps to provide content specialized to mobile constraints, such as limited bandwidth and the small screens and fiddly keyboards on many devices.

For client applications they tend to offer a message-based protocol. Some protocols are based on the ubiquitous HTTP web protocol. Others include audio and video content (using protocols such as RTP), and messages (using protocols such as RSS), etc.

Servers for browser applications need to provide content that meets the needs and limitations of the device's browser. They detect the device and browser by matching various protocol headers in the HTTP requests (e.g., the user-agent string, and endeavor to return appropriate content). The content may need to be pared down to meet limitations of size and complexity; and the format needs to match the markup language used by the browser (e.g., Wireless Markup Language [WML] for older phones).

1.2.5 Things That Do Not Quite Fit

Some technologies do not quite fit my classifications, for instance: Multimedia Messaging Service (MMS) is another messaging service and supported directly by many smartphones. However, unlike SMS it uses HTTP requests and responses (http://en.wikipedia.org/wiki/Multimedia_Messaging_Service) /

RSS "feeds" messages and updates from a server to clients who subscribe to the feed. RSS is similar to a browser application, but uses another piece of software to display the content. There are several interpretations of RSS (see http://en.wikipedia.org/wiki/Rss for more information).

1.3 CURRENTLY OUTSIDE THE SCOPE OF THIS BOOK

This book does *not* cover:

- Testing of the physical devices, their operating system, or of the platform (except where it affects applications that run on that platform);
- Automated test suites to certify the run time platform, such as Java Certification;
- Testing the internals of the base stations and carrier networks. However, we touch on these topics where they can materially affect the performance of mobile wireless applications; and
- Embedded devices or technologies I do not yet know about. However, many of the principles and techniques may be relevant.

If you would like to contribute ideas, experiences, and material please contact me—I would be happy to incorporate relevant work and acknowledge your contributions.

1.4 SCOPE OF MOBILE WIRELESS TEST AUTOMATION

My scope is fairly broad, it ranges from unit testing to system and field testing, and includes anything that helps to partially or fully automate testing of mobile wireless applications.

1.5 CHALLENGES IN TESTING MOBILE WIRELESS APPLICATIONS

We face various challenges inherent to testing mobile wireless applications ranging from practical limitations, to tedious, mundane tasks, to understanding what factors and issues affect the results of our testing (Figure 1.1).

Trying to test using all possible devices is impractical. Trying to multiply that testing across the rest of the factors exacerbates the problems (e.g., network operator, different versions of the underlying software, etc). Even the first stage of configuring handsets—so we can run and test an application—is error-prone and time-consuming.

FIGURE 1.1: Testing challenges.

The software installed on phones is constrained by various tradeoffs and decisions made by the provider of the phone. The provider may have customized the software installed on the phone to change the default behavior. Neither the original software nor the changes are well documented.

Detecting rendering, or display, issues generally needs someone to look at the content on the screen. As the User Interface (UI) code can be up to half of all the application's code, and as the UI is such an important factor for most mobile applications, the need for human involvement needs to be factored into much of our test automation.

Some factors that could affect the test results may be outside our direct control. They may hard to even identify and therefore even harder to measure. When we do test, accurate test data may be hard to obtain, and there are numerous gaps and contradictions in the data we have which we need to sift through to determine the key issues and their likely impact.

Measuring performance of mobile applications is an imperfect art, and particularly error-prone when trying to obtain consistent, accurate results.

1.6 PROBLEM SPACE

The world of mobile phones is complex; where the device is frequently provided by the network operator as part of a service. The software on these devices is often customized by the operator and the software may include significant changes to the user interface and to the functionality. For instance, some operators disabled the "Voice Over IP" (VoIP) feature from Nokia's very popular N95 handset to prevent their users from using this service.

There are hundreds of network operators—with multiple Internet price plans and particular price plans offer or prohibit particular services and have specific network configurations.

There are also hundreds of models of handsets, each may have many variants—e.g., firmware from a particular network operator—giving thousands of possible variations.

Combinations of price plan, network configuration, and phone firmware may limit or even disable part or all of an application.

The choice of handset also affects the runtime environment (e.g., some support Java ME and others support C++ programs). The preinstalled web browser(s) on a handset also determine which markup language(s) the server needs to use. While modern devices often support relatively complete xHTML or HTML markup, older devices might use more limited markup languages such as WML, C-HTML, or i-mode.

When testing application software we need to consider:

- The human languages (e.g., French, Kanji);
- The locales (e.g., UK English, Australian, and US English), which affect things like formatting numbers and currency symbols;
- Who pays for updates to be downloaded (users may be unwilling to pay to download updates OTA);
- How the software is installed on the device (e.g., in terms of security permissions); and
- The number of applications and versions you need to support in parallel.

Finally, there is the vital topic of what testing resources you have available to test each version and release of an application, and to decide how best to spend that time—e.g., which handsets from which network operators should you test with? We will cover the testing focus in more detail shortly.

The following diagram provides an overview of the problem domain.

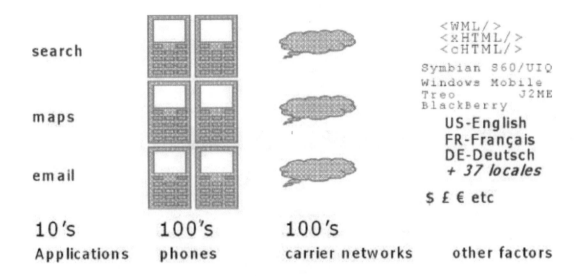

search

maps

email

```
<WML/>
<xHTML/>
<cHTML/>
```

Symbian S60/UIQ
Windows Mobile
Treo   J2ME
BlackBerry

US-English
FR-Français
DE-Deutsch
+ 37 locales

$ £ € etc

10's	100's	100's	
Applications	phones	carrier networks	other factors

1.6.1 Transcoding Web Content

Some content is unsuitable for devices—e.g., it may be too complex or contain images in a format not supported on a device. Google and other companies transcode content in order to make it more suitable for mobile devices. For example, Google's mobile search transcodes results by default for many mobile devices (and offer users the ability to view the original page if they prefer). Some carriers also transcode web content to do a similar job.

Here is a diagram of how a transcoder converts a graphical static web page to suit a generic web browser.

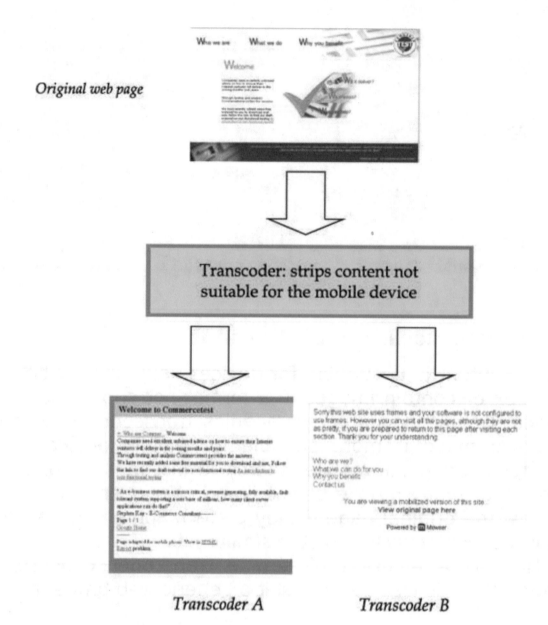

Original web page

Transcoder: strips content not suitable for the mobile device

Transcoder A   *Transcoder B*

Essentially a transcoder acts as an intermediate device which interprets both the HTTP requests and processes the HTTP responses. In the HTTP requests they typically process things like the user-agent string (to recognize which device is making the request), and in the responses they process things like content type and content length to determine whether content should be converted on-the-fly (dynamically) or whether it is appropriate to pass through unchanged.

1.7 OUR TESTING FOCUS

Given the vast problem space and our typically severely constrained resources, we need to focus our testing if we are to be effective. When automation is used appropriately we can be significantly more effective and reduce the overall time needed to test each software release. Some types of applications can be automated relatively easily and successfully, while others are more challenging (e.g., client applications). Finally some aspects are better tested manually—e.g., to assess the rendering of the UI on actual devices.

For applications that run on a range of devices, where there are lots of variations between devices and where upgrades can be expensive or difficult, we first want to focus on finding and addressing problems that would prevent users from being able to use the application on their device. These problems include:

- Finding incompatibilities ranging from not installing to poor rendering;
- Discovering and working around limitations in the software on the device, including browser issues, J2ME bugs, etc.; and
- Detecting content or behavior that may adversely affect the behavior of the device (e.g., where a large web page may not be shown at all on some devices).

Once we have tested for these issues the next step is determining whether users get the most technically suitable content for their device. For instance some smartphones from Nokia support both JavaME and C++, and they may have several web browsers installed (e.g., one that supports xHTML and another that supports HTML). C++ applications tend to be faster and take better advantage of the features of the smartphone, but they have to be "trusted" by both the user and often by the network operator who may prohibit unapproved software from being installed or used.

Pick a representative subset of the set of all the intended devices. I suggest you slice the set in various ways to increase the chances of finding meaningful bugs.

- Pick some of the most popular models and for these pick a model with the most popular version of the manufacturer's firmware (another term for the preinstalled software). For example, a Nokia N95, an iPhone with version 2.1 of the operating system, and a T-Mobile G1 with cupcake installed would represent a significant subset of the set devices with capable web browsers.
- Pick one model from a set of similar models—e.g., for the older Nokia Series 60 second edition devices a N6680 or an N70 are good representatives for the rest of the range of models. They have similar web browsers, JavaME runtime, and support the same C++ applications.

For all our applications we want our users to *like* using them. After all, unless we have a monopoly (e.g., for internal company applications), then our users have plenty of alternatives available. Here we focus on:

- Usability, the wow factor, etc.; and
- Performance, which is an umbrella term that includes: a user's perception of responsiveness, client-side rendering, OTA transmission times, and server-side timings.

Test design helps us to increase the effectiveness of each test, and the test coverage, without testing every possible permutation! Thankfully we can adopt existing techniques and good practices from elsewhere in the software testing communities. For example, we can use combination testing techniques to select our test cases and use exploratory testing techniques to help guide our testing.

1.8 OUR GOALS WHEN TESTING

Testing our software is a "means to an end"—part of the journey rather than the ultimate goal. However, if we have clear, measurable goals then we can keep track of how well we are doing and whether our testing is useful for the applications we are testing.

Here are some of the goals I have used over the years to help you identify goals that suit you and your work.

- *To ensure we deliver attractive, easy-to-use, working applications for as many users as practical.* Lots of happy, frequent users help show our software is successful and useful.
- *To have justified confidence in the quality of our software.* Providing accurate information on the quality of software is an important aspect of software testing. When we test well, and communicate the results so other people understand the strengths, weaknesses, risks, etc., with releasing our application, there should be few surprises after deploying the software to our users. Ideally, most of the bugs would be found and fixed before the software is widely used.
- *Fast feedback to developers.* Fast feedback helps them to fix the code while it is still "warm," while they are still intimately familiar with it.

- *To quickly detect issues so they can be addressed.* This is particularly relevant when the problem is related to external factors (e.g., an operator's network configuration or a specific handset model). Note: We tend to make changes to our software, as that is the fastest way to fix the issue from the user's perspective. We can then work with the relevant third parties to address the underlying issues in a more considered fashion.

For each of your goals, try to find ways to collect useful metrics (e.g., the number of bugs found in testing compared to the number reported by users). Dorothy Graham coined the term Defect Detection Percentage to measure these bugs—more information is available on her blog, http://dorothygraham.blogspot.com/.

1.9 OUR OVERALL TESTING STRATEGY

Start by underpinning manual testing with automated testing. As manual testing is very time-consuming, and often limited to testing through the limited user-interface of the mobile device, find ways to automate parts of the application code and the system. For example, use automated unit tests to test the business logic and the communications libraries of the client application. And automate the testing of the client-server protocols and interfaces by using custom clients, or independent automated tests from desktop computers, to send messages and inspect the received messages.

Rely on existing test automation tools and libraries if they exist: e.g., J2MEUnit for Java ME applications, and WebDriver for testing AJAX applications. In some cases there is no suitable tool, so you may decide to create one if you have the time, skills and resources. For instance we ended up creating JInjector so we could automate system testing and generate code coverage for our Java ME applications, and the IPhoneDriver for WebDriver. Both these tools are open-source and available free of charge. We have found "open-sourcing" our tools to be useful both for us and for the wider testing communities. We get their feedback and support, and they are able to use and extend our work.

Sometimes problems can be isolated and fixed sooner by splitting the code apart—e.g., to replace the UI with a text equivalent (sometimes known as a "headless" version). While doing so may seem like extra work, generally the debugging tools are less sophisticated than when debugging server code. Also, once you have a headless version, the tests should be able to run without (much) human involvement, unlike testing through the UI.

Consider reducing problems to their essential details, to divide-and-conquer issues. Note: mobile client code may be less elegant than equivalent server code, partly owing to restrictions imposed by the development platform and libraries, and partly because developers want to optimize to reduce size and increase speed of the application. Consider testing the servers in isolation, testing by using protocol emulators, testing locally on the device, etc.

Automate more of the build and deployment processes in order to accelerate and streamline the testing. Another benefit is that automated processes help to reduce the risk of human error in the deployment.

Seek ways to automate more of the end-to-end on-device testing, both to reduce the need for manual testing and to help identify device-specific issues cost-effectively.

Seek also to provide effective test output to reduce the effort required to identify and address problems. As mobile applications may have very restricted reporting capabilities —e.g., when running within the Java ME "sandbox", where applications need express permission to write to the filesystem—consider writing the results from the client to a server using HTTP, MMS, or even SMS messages.

One of my strategies, and one reason this book was written, is to allow other teams and groups to automatically test *their* software so that I can then get out of the way and leave them to it!

Finally for this section, do not be afraid to seek some quick wins as well as trying to address longer-term automation goals. In terms of testing mobile applications, some seemingly simple tools can significantly improve our effectiveness. These tools include:

- User-agent capture tools;
- Using SMS messages to send test URLs and download links to devices;
- Screen-capture tools; and
- Using "contact-sheets" that collect many screenshots into a single display, which enable lots of screenshots to be reviewed quickly.

All these tools are covered later in this book. You are welcome to add to the list and tell me about your favorite tips and tools.

1.10 CORE CONCEPTS

There are some core concepts which underpin our approach to testing mobile wireless applications. Let us start with connectivity.

- GSM and CDMA networks provide support for data.
- Phones include a GPRS (etc.) modem which provides the underlying connection. Web browsers on the phones (or

applications that use the internet for communications) use the data connection.

- We can either use phones or dedicated modems (which are also known as data cards) to establish a similar connection for testing.
- This material concentrates on HTTP connections that underpin the majority of connections between mobile phones and servers.

To help you understand how HTTP is used for connectivity the first chapter on automation "Testing techniques for markup applications" starts at a relatively basic level and builds the test code in small discrete steps until a basic test is created for a search page. Here is an overview of what that chapter includes:

- Sending an HTTP request and receive the response;
- Analyzing the request and response;
- Device Emulation, starting by adding a user-agent setting and adding more HTTP headers until we manage to convince the server that our tests should be treated as that device; and
- HTTP + Device Emulation + Content Validation.

Subsequent chapters include code snippets to highlight specific aspects or topics of test automation. A mix of programming languages are used, typically the same as would be used to develop the respective application.

Terms such as emulate and emulation are used throughout this material. Emulation means our software pretends to be the real device (e.g., our tests can send information which the server uses to determine whether the request is from a mobile device). Emulators are also provided by software vendors which behave sufficiently closely to real devices to enable our applications to run on our computers.

CHAPTER 2

Markup Languages

Markup languages combine instructions with content. The instructions range from formatting, e.g., to display text in *italics* to links for images, other web content, etc. The instructions are identified with tags, characters with a particular meaning for a given markup language. All of the web-based markup languages use angle brackets < > to indicate markup, e.g., <bold>Hello</bold> indicates the word Hello should be rendered in **bold**. The ampersand is used to encode angle brackets, and several other characters, in the content, e.g., > represents the < character in the content.

The granddaddy of all web-based markup languages is HTML. When companies wanted to deliver web-like content to mobile phones 5 to 10 years ago, they realized HTML was ill-suited. WML was the first of the markup languages developed with mobile devices in mind. Subsequently, a number of variations and enhancements have been created and used to reflect the changing needs of the mobile market. These markup languages include:

- xHTML,
- cHTML, used by iMode,
- WML, used by WAP 1.x.

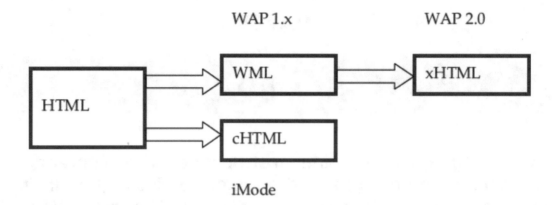

Two branches of mobile wireless markup languages from a common source. A more complete figure is available on Wikipedia
http://en.wikipedia.org/wiki/Wireless_Markup_Language.

Markup languages are rendered by software on mobile phones. There may be limitations or bugs in the rendering software. Conversely, the rendering software may need to account for flaws in the markup of a web page and cope with limitations of the hardware, e.g., in terms of screen size, fonts supported, memory limitations, etc. The net result is that the pages viewed by users on various phones may vary considerably in terms of appearance and even behavior—what works on one phone may not work on another.

2.1 EXAMPLES OF MARKUP LANGUAGES
Here is an example of an xHTML document, the home page of Google's mobile search in the UK (Figure 2.1).

```
<?xml version="1.0" encoding="UTF-8"?> (1)
<!DOCTYPE html PUBLIC "-//WAPFORUM//DTD XHTML Mobile 1.0//EN"
"http://www.wapforum.org/DTD/xhtml-mobile10.dtd"> (2)
<html xmlns="http://www.w3.org/1999/xhtml"> (3)
<head>
<meta http-equiv="Content-Type" content="application/xhtml+xml;
charset=UTF-8"/> (4)
<meta http-equiv="Cache-Control" content="no-cache"/>
<style type="text/css"> a:link {color:#0000cc;} a:visited
{color:#0000cc;} a:hover {color:#0000cc;} a:active {color:#0000cc;}
.c0 {vertical-align:middle;} .c1 {padding-bottom:1em;} .c2
{color:#ff0000;padding-right:2px;} </style>
<title> Google </title>
</head>
<body>
<div>
<img class="c0" src="/xhtml/images/google.gif" width="85"
height="35" alt="Google"/> (5)
</div>
<div class="c1">
<a href="/gmm?source=m&amp;dc=mobile-promotion" >Download Google
Maps</a> (6)
</div>
<form action="/m/search"> (7)
<div>
<input type="hidden" name="mrestrict" value="xhtml"/> (8)
<input type="text" name="q" size="15" maxlength="2048"/> <br/>
<input type="submit" name="btnG" value="Search"/> <br/>
<input type="radio" name="site" value="search" checked="checked"/>
Web<br/>
<input type="radio" name="site" value="images"/> Images<br/>
<input type="radio" name="site" value="local"/> Local listings<br/>
<input type="radio" name="site" value="mobile"/> Mobile Web (Beta)
<br/>
</div>
[... more content here, removed to fit the example on the page ...]
</form>
&copy;2007 Google <br/>
</body>
</html>
```

FIGURE 2.1: xHTML for Mobile Search homepage in UK.

Things to note in the xHTML document:

1. The document conforms to the XML specification;
2. The document type is XHTML Mobile 1.0. A link is provided to the online rules which apply to this document. The rules are defined in a Document Type Definition (DTD);
3. The XML namespace is referenced;
   Here is the equivalent Google home page in WML format (Figure 2.2).

```
<?xml version="1.0" encoding="UTF-8"?> (1)
<!DOCTYPE wml PUBLIC "-//WAPFORUM//DTD WML 1.2//EN"
"http://www.wapforum.org/DTD/wml12.dtd"> (2)
<wml> (3)
<head>
<meta http-equiv="Content-Type" content="text/vnd.wap.wml;
charset=UTF-8"/> (4)
<meta http-equiv="Cache-Control" content="no-cache"/>
</head>
<card title="Google"> (5)
<onevent type="onenterforward"> (6)
<refresh> <setvar name="q" value=""/></refresh>
</onevent>
<p> <b>Google</b> </p>
<p>
<input name="q" size="15" maxlength="255" emptyok="true"/> <br/>
Search: <br/>
<anchor title="search"> (7)
<go href="/m/search" method="get">
<postfield name="output" value="wml"/>
<postfield name="mrestrict" value="wml"/>
<postfield name="q" value="$(q)"/>
<postfield name="site" value="search"/>
</go>Web </anchor> <br/>
<anchor title="mobile">
<go href="/m/search" method="get">
<postfield name="output" value="wml"/>
<postfield name="mrestrict" value="wml"/>
<postfield name="q" value="$(q)"/>
<postfield name="site" value="mobile"/>
</go>
Mobile Web </anchor> <br/>
<anchor title="local">
<go href="/m/search" method="get">
<postfield name="output" value="wml"/>
<postfield name="mrestrict" value="wml"/>
<postfield name="q" value="$(q)"/>
<postfield name="site" value="local"/>
</go>
Local listings </anchor> </p>
<p>
&#169;2007 Google</p>
</card>
</wml>
```

FIGURE 2.2: WML for Mobile Search homepage in the UK.

4. The content type is set to application/xhtml+xml Setting the content type helps the browser to interpret the content correctly;

5. A link to the Google logo. The image is smaller and simpler than the one delivered to desktop web browsers such as Firefox;

6. Here is a link to promote Google Maps for mobile devices. This promotion only offered to mobile users, and is something we may want to test for;

7. When the user submits their search, the contents are passed to a mobile-specific version of Google search; and

8. A hidden field included. When the web browser sends the search request to the server it will include this field with

the search query. The server then knows the results should be returned in xHTML format.

Things to note in the WML document:

1. The document also conforms to the XML specification;
2. The document type is now WML 1.2. A link is provided to the online rules which apply to this document. The rules are defined in a Document Type Definition (DTD);
3. The content is within a pair of <wml> tags (the other is the last line of the example);
4. The content type is text/vnd.wap.wml which helps the browser to interpret the content correctly;
5. Content is displayed on individual "cards"; cards are a metaphor for the content, in a similar way that we describe web "pages." Cards can be grouped into a "deck";
6. "onevent" contains an action to be performed. Here the variable "q" is set to an empty string when the page loads in the phone's WML browser. "q" is the query parameter passed to Google's search engine; and
7. An anchor contains an action, here to perform a search by passing in the parameters, identified as "postfield", to the relative address on the server specified in the href parameter.

2.2 TESTING STRATEGY FOR MARKUP APPLICATIONS

Try to automate most of the testing for markup applications. Markup languages are designed to be processed by programs (such as the web browser) rather than being directly interpreted by a human. Therefore the markup languages are relatively simple to test using a program or script. Furthermore, there is often support for the web-based markup languages in many of the common programming languages, either directly or using software libraries, so we can take advantage of this support to reduce the amount of code we need to write.

Techniques such as matching patterns in responses are effective. The idea is simple: we expect certain things to be returned in the response, e.g., search results.

- People are good at instantly recognizing whether the results seem appropriate, one factor we seek is known patterns, e.g., a set of search results. However, we cannot predict exactly what each result will contain; instead we check the individual search results quickly and select any that seem relevant or interesting.
- In our automated tests we can articulate similar patterns in the response, e.g., that a "good" response may contain a number of search results. Each result will consist of a hyperlink, a snippet of text, and the source URL, etc. Again, it is unlikely we can predict the exact content of each search result.

We can therefore design our automated test to use patterns based on the hierarchy and/or the content of responses.

Two effective pattern-matching methods are: using regular expressions and using hierarchical path navigation (for instance with XPATH expressions). The examples later on provide examples of both methods.

2.3 EXAMPLE PROBLEMS

Here is a summary of typical problems that affect markup applications. Cookie and Transcoder issues may adversely affect other types of mobile wireless application and are generally "bad news" for those applications.

- Cookies are sometimes intercepted by a gateway, or a proxy server, provided by the carrier. Users may end up sharing a common cookie owing to the interactions between the web server and the gateway, or proxy server. These shared cookies can cause unexpected behavior, e.g., where one user's custom content is visible to other users who "share" a common cookie.

Automated tests may need to communicate across multiple distinct connections, in parallel, probably across several devices or computers, in order to trigger cookie sharing issues.

Theoretically tests from a single source could detect issues caused by other user's cookies affecting "our" use of the application. However, typically single source tests aren't designed to detect the problem, so—at best—they would detect a problem. But the problem is likely to be intermittent and treated as a "flaky test."

- Transcoders do not always help improve the mobile experience: we have noticed that they sometimes transcode data that needs to be returned as-is to our client applications, which prevents our applications from working correctly.

    We can write automated tests to determine whether a transcoder converts content inappropriately or stops content entirely. Problems with transcoders are generally resolved by experimenting with ways to set the headers so content is returned to the user correctly, or by contacting the company who is using or providing the transcoder.

- Devices from at least two manufacturers display a "413: Page Cannot Be Displayed" message for some web content. The error can be displayed under various conditions including factors such as:
    - The size of the HTTP response,
    - The complexity of the returned xHTML document, and
    - The length of the HTTP request (which could be over 500 bytes long in some cases).

  By designing server pages for typical issues the devices can be tested to determine their limits for these factors.

- Poor dropdown menu support (see below).

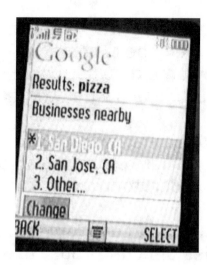

- Poor table support (see right).

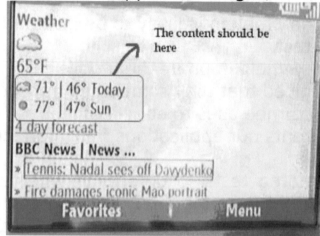

- Bold tag not supported in an anchor tag (see below).

CHAPTER 3

Testing Techniques for Markup Applications

In my experience the majority of automated testing for markup applications either uses HTTP requests and responses using a protocol library such as urllib or a browser-based automation library such as Selenium. Both approaches are useful and can be effective.

A web browser can be used as a simple way to quickly view the content returned for a given xHTML device. The browser will need an add-in to incorporate sufficient device headers to convince the web site that the request is from a particular mobile device. Similarly, WML can be rendered in Firefox by using another browser add-in. More information is available in the section titled "Utilities to help with testing browser applications."

3.1 GETTING STARTED WITH TEST AUTOMATION

Let us start with an overview of how you can implement your first automated script for testing browser applications:

- Implement commands to send a HTTP request and receive the response.
- Add some basic validation of the HTTP response in order to determine whether you are getting what you want, and not getting what you do not want:
  - Does the HTTP status code equals 200? (OK.)
  - Is the content type what you expected?
  - Is the content length appropriate? (You may be able to use a single exact value or specify a range of

acceptable values.)

- Add pattern-matching (e.g., to match the word "Results" followed by two numbers, followed by "of about") for the UK English search results on www.google.co.uk. See screenshot for an example of the search results page (Figure 3.1).
- Add a user-agent string that matches one from a specific mobile phone.
- If necessary, add other HTTP headers such as:
  - Accept,
  - X-WAP.

FIGURE 3.1: Google search results.

Consider adding code to detect failures, e.g., error messages that should *not* be in the responses. By detecting failures explicitly the tests will be more reliable (harder to fool) and errors can be handled sooner, rather than the test meekly waiting for the "expected result" until a timeout occurs.

3.2 EXAMPLES WRITTEN IN PYTHON

The examples here are written in Python, a flexible programming language that is easy to experiment with interactively as well as being able to handle large-scale programs. Several other programming languages are able to provide similar capabilities, so try yours if you don't find Python suitable.

```
>>> import urllib (1)
>>> request = urllib.FancyURLopener() (2)
>>> response = request.open("http://www.google.co.uk/m") (3)
>>> content = response.read() (4)
```

EXAMPLE 3.1: Four lines of Python to retrieve a web response.

With four lines of Python, shown in Example 3.1, we are able to retrieve the contents of Google's mobile search homepage for the UK site. *Note: the numbers of the bullet-points refer to the numbers in brackets from the preceding example Python code.*

1. import urllib does all the hard work of making the http request and returning the response, e.g., the library automatically handles HTTP redirection.
2. Create a request object, using one of the methods provided by urllib.
3. Make the request to retrieve the contents of the mobile search page, and assign the response to the response object.
4. The content of the response is returned using the read() method. We can now test the content directly by using standard Python functionality.

So with four simple lines of code we have the page content. We can also obtain more information, like the HTTP response headers by calling response.info()

```
>>> print response.info()
Set-Cookie: PREF=ID=dace3eeeb443a505:TM=1189341666:LM=1189341666:
S=HfvfxidI1xMPZ
930; expires=Sun, 17-Jan-2038 19:14:07 GMT; path=/; domain=.google.co.uk
Set-Cookie: MPRF=H4sIAAAAAAAAAMAAAAAAAAAAA; expires=Sun, 17-Jan-2038
19:14:07
GMT; path=/; domain=.google.co.uk
Pragma: no-cache
Cache-Control: no-cache
Content-Type: text/vnd.wap.wml; charset=UTF-8
Date: Sun, 09 Sep 2007 12:41:06 GMT
Server: GFE/1.3
Connection: Close
```

EXAMPLE 3.2: Display the HTTP response headers for WML.

And individual headers using response.info().typeheader

```
>>> response.info().typeheader
'text/vnd.wap.wml; charset=UTF-8'
```

EXAMPLE 3.3: Display the content type response header for WML.

Here you may notice something strange: the content type indicates the content is in WML, rather than xHTML or HTML. Google's search engine seems to be defaulting to the oldest markup language, rather than the most popular one—why is that?

Early mobile wireless devices used WAP requests and expected WML responses. Later phones were often able to support both WML and xHTML, while newer phones support xHTML and HTML. Since the search engine does not have any indication of what markup language the requester wanted, it sends the earliest version. Google search is able to deliver the correct markup provided it receives sufficient "clues" in the HTTP request.

Two key HTTP headers generally provide enough information for the search engine to deliver appropriate content for a given device. By adding these headers we should be able to "fool" the search engine into returning the content it would return to the equivalent phone model. Let us go through these headers one by one:

The first header to consider is the "Accept" header. This is part of the HTTP standard and is used by the requester to tell the server what types of content it is able to use. For a desktop web browser the Accept header may be set to

```
Accept: image/gif, image/x-xbitmap, image/jpeg, */*
```

This "accept" header tells the server the image file formats the browser can display, and then a catch-all for other content types, e.g., for text/html.

For a mobile browser that accepts xHTML, the Accept header will generally be set to:

```
Accept: application/xhtml+xml
```

As indicated with (1) in Example 3.4, in Python we can add this header using:

```
request.addheader('Accept', 'application/xhtml+xml')
```

```
>>> import urllib
>>> request = urllib.FancyURLopener()
>>> request.addheader('Accept', 'application/xhtml+xml') (1)
>>> response = request.open("http://www.google.co.uk/m")
>>> content = response.read()
```

EXAMPLE 3.4: Adding the accept header for xHTML.

```
>>> print response.info()
Set-Cookie: PREF=ID=f7be92230ff7006f:TM=1189343788:LM=1189343788:
S=NiHOJyU0dt5Sf
b6C; expires=Sun, 17-Jan-2038 19:14:07 GMT; path=/; domain=.google.co.uk
Set-Cookie: MPRF=H4sIAAAAAAAAAMAAAAAAAAAAA; expires=Sun, 17-Jan-2038
19:14:07
GMT; path=/; domain=.google.co.uk
Pragma: no-cache
Cache-Control: no-cache
Content-Type: application/xhtml+xml; charset=UTF-8
Date: Sun, 09 Sep 2007 13:16:28 GMT
Server: GFE/1.3
Connection: Close
```

EXAMPLE 3.5: Displaying the HTTP response headers for xHTML.

And using response.info().typeheader this time shows the response is in xHTML—phew!

```
>>> response.info().typeheader
'application/xhtml+xml; charset=UTF-8'
```

EXAMPLE 3.6: Display the content type for xHTML.

At the time this example was created, September 2007, Google was promoting the "Google Maps" mobile application to users who had suitable phones. I wanted to write a simple test that checks whether the link is available on the xHTML homepage. As the link is only offered to users with suitable phone models, we need our test to pretend it is one of those phones. The User-Agent HTTP header is what we need to achieve this.

Here is an example of a user-agent string from a Nokia 6230 phone:

```
'Nokia6230/2.0+(04.43)+Profile/MIDP-2.0+Configuration/CLDC-1.1+UP.
Link/6.3.0.0.0'
```

We can add this to our request by adding another header (shown in bold and indicated with (1) in Example 3.7:

```
request.addheader('User-Agent',
'Nokia6230/2.0+(04.43)+Profile/MIDP-2.0+Configuration/CLDC-1.1+UP.
Link/6.3.0.0.0')
```

```
>>> request = urllib.FancyURLopener()
>>> request.addheader('Accept', 'application/xhtml+xml')
>>> request.addheader('User-Agent',
 'Nokia6230/2.0+(04.43)+Profile/MIDP-2.0+Configuration/CLDC-
1.1+UP.Link/6.3.0.0.0') (1)
>>> response = request.open("http://www.google.co.uk/m")
>>> content = response.read()
```

EXAMPLE 3.7: Adding a user-agent string to emulate a Nokia 6230.

3.2.1 A Test to Detect if Google Maps Is offered to Mobile Users

Let us start by using a simple search to find out whether Google Maps is mentioned anywhere on the page.

```
>>> content.find('Google Maps')
807
```

EXAMPLE 3.8: Using string search to find out if the content contains Google Maps.

If the string is not found content.find() returns -1.

```
>>> content.find('Google Mapsx')
-1
```

EXAMPLE 3.9: The String search returns -1 if the string is not found.

So, a simple string search is enough to get us started, and here indicates that the string "Google Maps" is contained in the returned content. We could now write more involved "string searches," e.g., to find and extract the URL for the download and make sure the page does not have multiple links to Google Maps, etc. However, Python offers a better approach using the very powerful regular-expression tools.

3.2.2 Using Regular Expressions in Our Test

Four more lines of Python code are enough to perform the regular expression match and return the link to the download page.

```
>>> import re (1)
>>> rx = re.compile('<a href.* Google Maps') (2)
>>> m = rx.search(content) (3)
>>> m.group() (4)
'<a href="/gmm?source=m&amp;dc=mobile-promotion" >Download Google Maps'
```

EXAMPLE 3.10: Using a regular expression to get the download link for Google Maps.

Here are the four steps required:

1. Import the re regular expression module (provided as a standard Python library module).
2. Define and compile the regular expression. The .* characters mean: match any characters between the href and the string "Google Maps".
3. Perform the search on the content of the web results. If matches are found the m variable will be assigned to point to the set of matches.
4. m.group() returns all the matches, here there is only one match, which is displayed on the next line. The link is a relative link: /gmm followed by some parameters that are useful for tracking the promotion.

We could easily refine this code to extract, and even download, the Google Maps software.

3.2.3 Combining XML With Regular Expressions

Both WML and xHTML return XML documents; therefore we can use XML processing techniques to locate content of interest. The following example demonstrates how to use standard xml Python modules to match a string or even a regular expression and return the link if it exists.

```python
import re
import sys
from xml.dom import minidom

def getLinkFromXhtml(content, text_regex):
  """getLink returns the href link for a given text_label.

  Args:
    content: the source content e.g., an xHTML response.
    text_regex: the text to match as a regluar experession.
  Returns:
    The href if the test is found, else None.
  """

  doc = minidom.parseString(content)
  links = doc.getElementsByTagName('a')

  rx = re.compile(text_regex)

  for i in links:
    if i.hasAttribute('href'):
      t = i.firstChild
      text = ""
      while t:
        if t.nodeType == t.TEXT_NODE:
          text += t.data
        t = t.nextSibling
      match = rx.search(text)
      if match:
        return str(i.toxml())

  return None
```

EXAMPLE 3.11: getlink.py.

3.2.4 Using XPATH in Our Tests

There is a powerful free Python module called Amara that can be used to address the xml structure using an intuitive dot addressing, e.g., html.body.div. Amara also supports XPATH expressions, e.g., //div[0].

Here is an example of using Amara to test whether the Google Maps link is available. The link is returned if Google Maps is found in the text of a link.

```
import re
import amara

doc = amara.parse(open("mobile-homepage.xhtml"))  (1)

def getHrefFromXML(doc, search_regex):  (2)
    """Returns the href link if the in search_regex is found in any
<div> tags.

    Assumes the links are in the html body's div tags.

    Args:
      doc: an amara xml object
      search_regex: the regular expression to match in the href text

    Returns:
      the href as a string if the pattern is found, else None.
    """
    rul = re.compile(search_regex)
    for item in doc.html.body.div:
        try:
          # print str(item.a.xml_children[0])
          # print type(item.a.xml_children[0])  (3)
          p = rul.search(item.a.xml_children[0])  (4)
          if p:
            return item.a.href  (5)
        except:
          pass

    return None

if __name__ == "__main__":  (6)
  print "should return: 'u/gmm?source=m&dc=mobile-promotion'"
  print getHrefFromXML(doc, "Google Maps")  (7)
```

EXAMPLE 3.12: Using Amara to extract the Google Maps link.

Example 3.12 is a little more involved, but still relatively easy to follow:

1. Use Amara to parse the web result. In this example, the code uses a saved example of the search results. To read the result from the web, simply replace the filename with the full URL starting with http://
2. getHrefFromXML is a helper method, which we can use many times to match various regular expressions.
3. These two lines are commented out (using the # character). These are examples of debugging the helper method and these lines are generally removed once we have debugged the method.
4. By using item.a.xml_children[0] we restrict the match to one part of the XML structure in the response.
5. Return the match, if found. Otherwise the helper method will return None (a Python reserved word that we use to indicate no match was found).
6. This is a standard convention in Python to execute the subsequent code *iff* the script is being run directly (rather than as part of a library).
7. Test the helper function by searching for "Google Maps".

3.3 SUMMARY OF THE EXAMPLES IN PYTHON

These sample Python scripts have demonstrated the various ways we can process xHTML content to:

- Make a basic HTTP request.
- Set the content-type to control the markup language returned by the web server.
- Set the user-agent to emulate a particular phone model.
- Parse the returned xHTML content to determine whether the Google Maps for Mobile link is provided for particular

phone models.

3.4 BUILDING ON YOUR FIRST AUTOMATED SCRIPTS

We can build on these basic scripts to automate more of our mobile browser testing.

3.4.1 Data-Driven Tests

We can make the tests "data-driven" as follows:

- Define a of input parameters. Consider parameters such as:
  - User-agent string,
  - Accept header,
  - URL,
  - Input parameters,
  - HTTP verb (GET or POST), and
  - Expected results, e.g., should there be a link for 'Google Maps'? There may be specific variations for particular devices, e.g., different download links.
- Read these from an external source.
- Return the results, e.g., as a file for later checking. Some basic checks, e.g., for the contenttype may be performed by the program.

3.4.2 Obtaining Metadata to Drive Our Tests

Metadata facilitates our automated tests, e.g., user-agent strings, accept headers, etc. Data sources include:

- commercial,
- WURFL,
- local data, and
- http://www.pycopia.net/webtools/headers (which captures the headers from the web browser on the device you are using and can email them to you, to save copying and pasting).

Find ways to rate and weigh the data quality, as data is often incomplete, contradictory, or wrong. You may decide it's worth filtering and merging useful data into a common pool.

3.4.3 Using Metadata

Once you have collected, filtered and merged your metadata it can be made available as a common source of data for your tests (and possibly also for your production systems). Figure 3.2 illustrates a single example of metadata for a Nokia 6230i device.

```
deviceFile = open("metadata.csv")
for line in deviceFile:
    (model, userAgent, accept, …) =
        line.split(',')
```

Do stuff with device data…

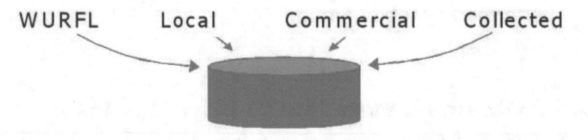

MODEL	USER-AGENT	ACCEPT	SCREEN_WIDTH	SCREEN_HEIGHT
Nokia 6230i	Nokia 6230i/2.0…	Xhtml	208	208

FIGURE 3.2: Combining sources of data with an example of data for a phone.

3.4.4 Test Using Carrier Networks

Carrier networks can sometimes affect the results returned, and sometimes even corrupt the content. If you have a suitable data connection (e.g., using a GPRS modem or a phone connected as a data modem) you should be able to run the tests over carrier networks and compare the responses returned over each carrier. Generally we expect all the responses to be identical (for the patterns and content we expect to receive). If differences occur, potentially, a given carrier's network/infrastructure has modified the content. The differences should then be investigated and assessed in terms of the impact they have on users.

Your test scripts should:

- Record the carrier network used (this may be as simple as using a user-specified string, or more complex, e.g., by querying the modem using on of the common GPRS AT commands : AT+COPS? for the currently connected carrier network before running the tests);
- Compare the response against a known reference page (possibly downloaded over your wired network); and
- Highlight differences, e.g., using a diff program for a user to compare the differences. Of course you can write more sophisticated comparison algorithms to reduce the need for people to get involved.

Note: One of the appendices includes information on how to configure a suitable data connection in Linux.

Networks and data plans vary and can affect the functioning of your applications. While you could test using an application on a phone, the problems may be hard to isolate from other issues related to the phone(s) you are using to test. A good way to isolate carrier and data plan issues is to run a custom test program on a computer with a wireless data modem, as Figure 3.3 shows.

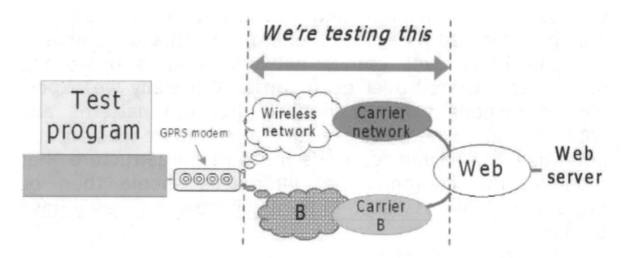

FIGURE 3.3: Testing the carrier network.

3.4.5 Timing the Request/Response Pair

One of the key frustrations of users for mobile wireless applications is how slow they are. Virtually all modern programming languages offer the ability to time how long things take to happen. If you start a timer just before making the HTTP request and stop it as soon as the HTTP response has been received, you will have a good idea of how long the OTA part of the transaction takes. If you choose to run similar tests for a range of carriers you can obtain a rough idea of the speed of each carrier for your local network conditions. Note: the speed varies significantly as local conditions change. Factors include:

- The protocol used to connect (e.g., GPRS is generally much slower than 3G),

- The number of active connections to the operator's local base station,
- The distance from the base station,
- The physical topology,
- The performance of the mobile wireless device, and
- The weather.

Measuring the end-user experience is much more involved and outside the scope of the current material.

3.4.6 Implementing Rule-Based Tests

In our experience, certain patterns of xHTML cause problems for particular phone models. Sometimes the issue is size of the response, for others tags such as the bold tag cause the contents to be rendered poorly, some handsets have limitations on the image formats they can handle, one phone even makes the text larger when the tag specifies the font size should be small, etc.

The effect ranges from the page not being displayed at all to minor rendering issues.

We do not want our users to have a poor user-experience, so we want to ensure our content will not trigger issues on their phones. At one point we created individual tests to detect whether particular content was appropriate for a given model of phone. However, that work was time-consuming and did not scale well. Therefore we designed and implemented a rule-based engine that can query URLs while emulating a wide range of phone models. The responses are then checked using rules which detect any violation of the known issues for that model of phone.

We gather issues from a variety of sources, e.g., from bug reports, server logs, manual testing, etc; identify them (in terms of which devices are affected and when the issue occurs); quantify them (in terms of severity, impact, likelihood, etc.); and write rules to detect the issue.

We simulate the affected devices using a mobile device database (loosely based on WURFL, an open-source effort to collect mobile device characteristics). We make requests to a variety of web sites that should work for mobile devices and apply the rules to the response received. If a rule is broken the software reports an error, i.e., if the content would cause an issue for the emulated device.

Generally we are able to manually verify whether a bug really occurs by testing with a real phone. This helps us to remove false positives from our system.

Our developers are then able to modify the application software to prevent the issue from occurring. Their changes are generally also tested manually, e.g., for aesthetics and to ensure the content renders correctly on the affected devices.

3.4.7 Probe Servers

Sometimes we are left with a number of open questions, such as:

- What are the capabilities various devices?
- What content causes problems?
- What is the signature of each device?

One way to obtain the answers is to create a "Probe server" that interacts with devices to obtain the answers we need. The tests need to be unambiguous for the tester, particularly in terms of their ability to provide accurate answers. If the tester is confused on what the "correct" answer is then the tests are likely to take longer while the quality of the data may also be compromised. Also, try to reduce the time required for each test. I am aware of one commercial vendor that requires a tester to manually execute more than 1,000 distinct tests per device in order to identify the device's characteristics, which take several days to execute. I don't envy their testers!

3.4.8 Strengths and Weaknesses of Rule-Based Testing

Strengths:

- Great for regression testing and for testing new applications that need to deliver content to mobile devices.
- Rules are generally relatively easy to codify, and easy to execute automatically.
- Additional rules can be added to check for accessibility using internationally recognized guidelines from the Web Content Accessibility Guidelines (WCAG, http://www.w3.org/TR/WAI-WEBCONTENT/).
- It can test some aspects of user experience, e.g., to detect consistency across multiple web sites for a large companies. Potentially, CSS can be checked dynamically (as it can be affected by client-side scripting, the model of web-browser, etc.).

Weaknesses:

- Relies on the quality of the mobile device database, which has proven to be inaccurate.
- Each issue needs to be identified, quantified and coded. The coding tends to require some technical

understanding of the underlying markup language, regular expressions and/or XPATHs, and Java.
- False positives need to be tested manually (and false negatives need to be fixed too).
- Does not currently simulate user-input.
- Does not test navigation, or scripting (e.g., JavaScript).

3.4.9 A Complementary Tool to Rule-Based Tests

One way to learn about the characteristics and issues for mobile phones is to use the phone to interact with a known set of web pages that contain various test cases, e.g., to determine which images are supported. If the web pages are interactive, then the user can provide feedback online while executing the tests (e.g., to confirm whether an image has been displayed correctly or not). The resulting data can then be codified into rules for that model of phone.

Furthermore, by running the rule-based checker against the test web pages the accuracy of the rule-based checker can be verified.

3.4.10 Is Appropriate Content Being Served?

Large, sophisticated international web sites need to deliver content that is appropriate for each user. Factors that affect the selection of content include things like:

- The device being used (e.g., an xHTML phone, an iPhone, a desktop browser) and its capabilities to display content.
- Whether scripting is supported on the client?
- The location of the user (which can affect the selection of localized content and the human language returned).

- User preferences, which may override default language and content selections.

Sometimes we redirect an initial request to a more suitable user interface, e.g., from the general search to the mobile search, based on the capabilities of the device.

Automated tests can be used to test all aspects of these requirements to varying degrees of accuracy. For instance the device being used can be emulated quite easily using HTTP headers. Implementing support in our tests to process scripting support is significantly harder. If your tests need to process scripts, consider using third-party open-source libraries such as HtmlUnit.

Location can be inferred from data such as source IP address and other factors such as phone-specific HTTP headers.

User preferences are often stored in cookies and cookie content can be added relatively easily to HTTP requests. Some more sophisticated cookie strategies require more involved programming to mimic the behavior of the way a device handles them.

3.5 TIPS WHEN IMPLEMENTING AUTOMATED TEST SCRIPTS

- Pick a simple web page that does not change, consider running a local web server to serve test pages initially.
- Use a network traffic analyzer to record the requests and responses.
- Get a trustworthy HTTP protocol reference (see Appendix A for some examples).
- Try to download pages from a web server that captures the HTTP headers. Access that web server from several

different mobile phones and save the captured data (see Appendix A for an example of such a server).

3.6 TEST TOOLS FOR BROWSER-BASED APPLICATIONS

3.6.1 Using Web-Testing Tools

For xHTML, the content is sufficiently similar to HTML to allow us to use many of the generic automated test tools, provided the tools offer the ability to set and read the HTTP request and response headers.

HttpUnit and HtmlUnit enable Java developers familiar with the JUnit test framework to create similar tests as easily as using Python (except Java does not provide an interactive development capability). HtmlUnit is more capable and, as mentioned earlier, includes support for testing the JavaScript scripting language.

3.6.2 "Mobile Readiness" Tools

Various software tools and web sites provide the ability to test the "readiness" of a web site for mobile wireless devices. For instance, http://ready.mobi/launch.jsp?locale=en_EN allows a web site to be checked online in terms of readiness for mobile devices. There are other similar sites and services available; however, this one has more of a testing focus. It implements the w3c mobileOK basic tests http://www.w3.org/TR/mobileOK-basic10-tests/ and is able to simulate a variety of common handsets.

3.6.3 Utilities to Help With Testing Browser Applications

Browser add-ons such as: wmlbrowser, web developer, user agent switcher, and modify headers (all for Firefox) make manual testing significantly easier. All of these add-ons can be installed directly from Firefox from the tools menu. On Microsoft Windows the menu option is "Extensions" on Linux it is "Add-ons."

XML tools, such as Oxygen (http://www.oxygenxml.com/), a commercial product, reduce the challenges of working with XML.

Firebug is a stunningly useful extension for Firefox that provides incredible analysis and debugging tools. Features range from displaying the XPATH of an element when the mouse is "hovered-over" page elements on an xHTML web page, to interactive debugging of JavaScript.

Another important tool for testing browser applications is a network traffic analyzer that helps decode the requests and responses actually sent between your machine and the server, rather than what you *think* or *hope* is being sent.

CHAPTER 4

AJAX Mobile Applications

AJAX applications have been around for several years for desktop web browsers. For instance, many of the popular web-based email applications make extensive use of AJAX. AJAX (Asynchronous JavaScript and XML) is an umbrella term for rich browser-based applications. They run much of the UI interactions and some of the data processing within the web browser on the client device. Smartphones, including iPhones and ones that use the Android software platform, are capable of running these applications, and lots of AJAX applications have been developed specifically for these devices. More information on AJAX is available on the web, including http://en.wikipedia.org/wiki/AJAX.

AJAX applications make extensive use of JavaScript, which repeatedly changes the content within the phone's web browser by manipulating the Document Object Model (DOM). The JavaScript also requests data from the web server and processes the data, e.g., incoming chat messages and status changes for the Google Talk application (Figure 4.1). AJAX applications also make requests and receive responses asynchronously. The data can be formatted in various ways; two popular formats are XML and JavaScript Object Notation (JSON). XML was introduced in Chapter 2, JSON is a more compact data structure—more information is freely available from http://www.json.org/.

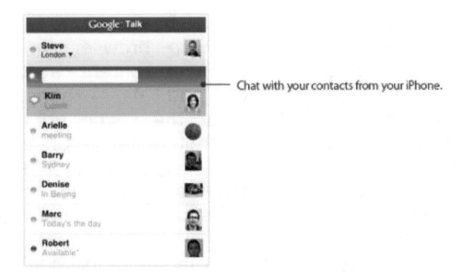

Chat with your contacts from your iPhone.

FIGURE 4.1: Example of an AJAX mobile application: Google Talk for the iPhone.

FIGURE 4.2: Image-bundle from Google Tasks.

When AJAX applications are developed for mobile devices, the priorities need to change, e.g., animation is less important, while application size and the design of the client-server protocol are more important owing to the constraints imposed by the device. Also connections are more likely be intermittent, e.g., when using the device on the move, so the application may need to be self-sufficient and be able to preserve updates made in the client application until network coverage is available again.

Like desktop web browsers, devices such as the iPhone include a cache within the web browser for content downloaded from servers. When web content can be stored in the cache, performance and user-experience are significantly improved. However the cache on a mobile device is tiny in comparison to the cache of a desktop browser; so canny application designers will tweak their implementation and experiment with actual devices to find ways to maximize the use of the cache. A useful article on the topic is available online (http://yuiblog.com/blog/2008/02/06/iphone-cacheability/).

AJAX toolkits such as Google's Web Toolkit (GWT) provide image-bundles that combine several small, similarly sized images into a single file (see Figure 4.2 for an example from Google Tasks) and programmatically show the correct picture.

The image-bundle from Google Tasks is only 1.4 KB so fits comfortably into the iPhone's browser cache.

AJAX, and AJAX frameworks in particular, make extensive use of JavaScript, which can be quite verbose (and therefore slow to download and less likely to fit in the browser's cache). The JavaScript can be compressed, and may also be obfuscated, which tends to reduce the size of the file. Obfuscated code is harder to understand or debug, which can make creating automated tests more difficult.

The designers of mobile devices may provide custom features in the web browser, e.g., so web applications can detect the device being rotated from portrait to landscape. The developer may choose to include conditional logic to detect and support custom features for particular devices, or create specific versions of their code tailored for that device. (Currently Google Tasks has distinct versions for iPhone, Android, and a generic less capable XHTML version for other phones.)

4.1 TESTING CHALLENGES FOR AJAX MOBILE APPLICATIONS

Few mobile devices support test automation directly, and we need to be able to test rich UI interactions of the application running within a web browser. The behavior of the applications is asynchronous, e.g., for an email application new emails may arrive from time to time from the server without any action by the user. The application can consist of large, compressed and obfuscated bundles of JavaScript that pushes the browser to the limits of what is possible. And, as so often happens, there are often differences in the features, capabilities, and behaviors of the browsers between devices.

Testing how well an application copes with intermittent network connectivity can be convoluted and may require an intermediate device, e.g., to intercept and modify network packets.

If desktop web browsers are used, e.g., with browser automation software such as WebDriver (see next chapter for more information), there will be significant differences in the capabilities and characteristics of that browser compared to the ones used in the mobile device, e.g., in terms of performance. Also, the desktop browsers are unlikely to support or provide the custom features such as rotation events.

4.2 EXAMPLES OF PROBLEMS WITH MOBILE AJAX APPLICATIONS

Mobile AJAX applications are relatively new, the support and behavior varies from one version of a browser to another. Developers want to take advantage of mobile-specific features, e.g., screen rotation and touch interfaces which complicate the client-side application logic. Here are some examples of problems seen with mobile AJAX applications:

- Coping with extensive DOM manipulation performed by the JavaScript. The DOM is one representation of the content in the web browser. During our testing of a recent application we noticed that one model of phone displayed a black box over part or all of the screen while the DOM manipulation was happening—not very attractive for the user.
- Complex DOM structures, with deeply nested structures, lacking hooks for testability (css classes and html ids can make testing much easier and robust).
- Duplicate, missing, and out-of-order content because of flaws in the asynchronous data processing.
- Applications not detecting or coping adequately with being suspended, losing the connection, etc.

Let us move on to how we can automate some of the testing for mobile AJAX devices.

CHAPTER 5

Testing Mobile AJAX Applications

We should be able to automate many aspects of testing Mobile AJAX applications. We have several options in terms of automation, including: using desktop test automation tools; using embedded browsers; and even automating some of the more complete device emulators, such as the iPhone emulator included in Apple's development tools.

5.1 USING DESKTOP BROWSER AUTOMATION TOOLS

Desktop browser automation tools have a long pedigree and there are tens of tools available, both commercial and free-of-charge. I'll cover two free open-source desktop browser tools: Selenium (http://selenium.openqa.org/), and WebDriver (http://code.google.com/p/webdriver/).

5.1.1 Selenium

Selenium is available in several guises: your choice will depend on things like your programming skills. Selenium includes a simple interactive development environment which can also record your interaction with a web site. The underlying scripts are stored in HTML tables, and the language is called Selanese. Selanese can also be installed on the same web server as the application to be tested, this is called Selenium Core. And finally there is a client-server version which supports a range of programming languages.

A subset of Selenium Core can be used to run on some devices, e.g., on the iPhone; however, in practice it is not very appropriate for even moderately complex AJAX applications. More practically, Selenium RC can be used in a desktop web browser with HTTP header emulation. The Selenium web site includes information on how to decide which version of Selenium to use.

5.1.2 WebDriver

WebDriver is more capable and powerful and is designed to support the needs of programmers. The primary programming language is Java, and the API enables programmers to work quickly and efficiently in their integrated development environment (IDE), e.g., Eclipse or IntelliJ. Python bindings are also available, and other languages may be supported in future.

WebDriver is designed so it can use a remote WebDriver "server." They communicate using a JSON protocol (http://code.google.com/p/webdriver/wiki/JsonWireProtocol) and a server has been implemented that runs on an iPhone, with another expected for the Android platform. There are some limitations when using these "on-device" servers, e.g., they may share the communications channel (3G or WiFi), they use a webview component[1] rather than the main web browser, and currently only a subset of the features are supported. However, they show promise, and as all the code is freely available to modify without charge people can tailor the code to suit their test automation needs, with the option to contribute their work back into the open-source community.

WebDriver and Selenium are going to be integrated. WebDriver overcomes several limitations caused by Selenium's implementation as JavaScript running the browser; and WebDriver provides a cleaner programming API which helps to reduce the size of the tests. In turn,

Selenium supports more web browsers and platforms, so it extends the range of WebDriver to otherwise unsupported combinations of browser and platform—albeit with some technical limitations.

I would recommend using WebDriver unless you have a compelling need to run the tests on an unsupported platform, e.g., in the Safari web browser, or using an unsupported programming language e.g., Visual Basic.

5.1.3 Customizing Desktop Web Browsers

For mobile AJAX applications we can combine these desktop web browser automation tools to send custom HTTP headers, e.g., for the user-agent string.

- Safari includes support for custom user-agent strings directly from the developer menu.
- Firefox has an add-in available which provides similar functionality.
- Microsoft's Internet Explorer's user-agent can be customized using registry settings, see http://www.pctools.com/guides/registry/detail/799/.

5.1.4 Limitations of Using Desktop Web Browsers

While desktop web browsers are incredibly useful to automate testing of mobile AJAX applications, they behave differently from the mobile browsers and inherently some important factors we want to test remain elusive or impractical to test.

The browser on a mobile device tends to use a fixed area covering most of the screen, although the iPhone offers a "full-screen" mode (available with version 2.1 of the iPhone software, http://ajaxian.com/archives/iphone-full-screen-webapps). On-screen keyboards appear within the fixed area when the user wants to input text, and disappears afterwards. Mobile phone web browsers do not support (much) UI customization so programmers learn how to design their application so it "fits" the limited screen area exactly. Desktop browsers, in contrast, can be resized and support customization and extensions so that the screen dimensions vary significantly. While each desktop browser can be sized to roughly match the dimensions of a device's browser, it is hard to control the browser's settings programmatically.

Also, differences in how the browsers lay out and render content means the UI of an application in a desktop browser differs significantly from the layout on the actual device. Testing the appearance of the application is therefore hard to do and the results may be unreliable.

The performance of a desktop computer, and therefore also the browser, is vastly different from that offered by a mobile device, e.g., in terms of the number of HTTP requests the browser sends in parallel, and the page caching (covered earlier). In short, do not rely on timings measured using desktop computers and web browsers when trying to test your application.

Sometimes the mobile browser uses JavaScript features which are missing from the mainstream web browsers. Some features are custom to a device [e.g., the rotation events on an iPhone], while others will be supported [e.g., querySelectorAll()] but are not available yet [e.g., Firefox 3.0 does not include it, version 3.5 will]. I noticed this when writing tests for Google Tasks when I changed the user-agent from an older version to emulating a 2.2.1 device and the application stopped running in Firefox 3.0. Thankfully, Firebug quickly enabled us to identify the problem, as the following figure shows (Figure 5.1).

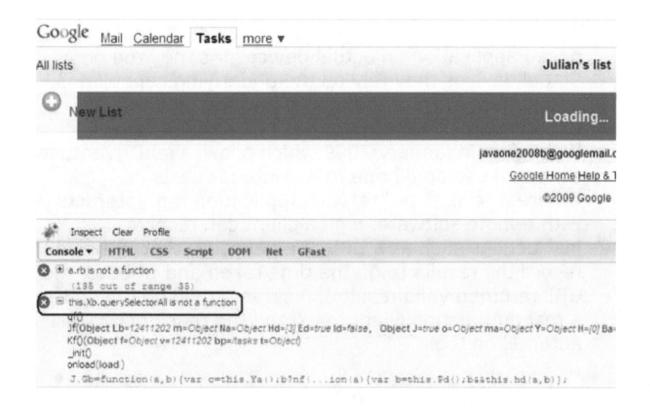

FIGURE 5.1: Firebug reports a missing JavaScript method.

There are several workarounds, including: upgrading the browser to one that does support the JavaScript method, limiting our set of user-agents to those that do not cause the application to use the newer methods, or for the ambitious—implementing and injecting the missing JavaScript methods. Note: Google Tasks includes a version of querySelectorAll within the page when serving older iPhone browsers.

5.2 USING AN EMBEDDED BROWSER

High-end devices, such as the iPhone, may include an embedded web browser component (for the iPhone it is called UIWebView). Software can be written to use these embedded browser s for test automation. Here are some suggestions on how they can be used:

- To test the latency and general behavior of a browser-based application on actual devices. As they run on actual devices they rely on the underlying capabilities and performance of the device and the network. An experimental iPhoneDriver has been added to WebDriver in January 2009 which allows WebDriver test scripts to use an iPhone to execute requests.
- Create a "client" or "server" application that interacts with remote software, e.g., a client can receive instructions such as a URL, then execute that URL and record the results (e.g., the time taken and whether the URL returned valid results). A server could be driven by a test automation client to extend the reach of the test automation tool.

5.3 USING SIMULATORS

Simulators are available for many of the high-end devices, often as part of a development toolkit. These can be used to make manual testing easier and provide a way to do testing when devices are not available. In theory the simulator could be automated using desktop application automation tools that operate at the Graphical User Interface (GUI) level. I would not recommend automation of simulators as a general solution unless the simulator provides good support for test automation. Otherwise the effort is likely to be substantial, and testing with actual devices will provide more faithful and complete results.

5.4 GENERAL TIPS

The development team can make testing much easier and more robust by adding custom, unique identifiers for key elements (e.g., for a search box), which makes the test scripts much easier to write and reduces the likelihood of your tests breaking each time the code is updated. The identifiers are implemented using the ID attribute in HTML.

Neither Selenium nor WebDriver provide a way to directly detect activities or actions in the AJAX application. Therefore our automated tests have to rely on polling the web browser in order to determine when the UI has changed.

> "Note also that the purpose of the automated testing is to increase confidence that the application works as expected; even if the emulation isn't perfect, it means that there's less manual testing required, which can only be a Good Thing."
>
> —Simon Stewart, author of WebDriver, July 2008

5.4.1 Selenium Tips for Mobile AJAX Automation

- Selenium IDE is available for Firefox. Create a custom Firefox profile that includes the user agent add-in. Add user-agent strings for the mobile device(s) you want to test. Then record a session in the IDE to create a basic script which can be incorporated into your automated tests.

5.4.2 WebDriver Tips for Mobile AJAX Automation

- When testing with a desktop web browser, use Firefox and the FirefoxDriver. Firefox is easy to customize and extend, and the WebDriver interface is the most capable as a result.
- In the test programmatically create a custom profile which sets sufficient HTTP headers to enable the browser to convince the server our requests have been sent from the device we want to emulate. See the following code example to configure Firefox to emulate an iPhone device.

```
private static final String IPHONE_USER_AGENT_V1_1 =
    "Mozilla/5.0 (iPhone; U; CPU like Mac OS X; en) AppleWebKit/420.1 "
    + "(KHTML; like Gecko) Version/3.0 Mobile/3B48b Safari/419.3";

/**
 * Returns the WebDriver instance with settings to emulate an iPhone V1.1
 */
public static WebDriver createWebDriverForIPhoneV1_1() {

    FirefoxProfile profile = new FirefoxProfile();

    // blank headers th at would otherwise confuse the web server.
    profile.setPreference("general.appversion.override", "");
    profile.setPreference("general.description.override", "");
    profile.setPreference("general.platform.override", "");
    profile.setPreference("general.vendor.override", "");
```

```
    profile.setPreference("general.vendorsub.override", "");

    profile.setPreference("general.appname.override", "iPhone");
    profile.setPreference(
        "general.useragent.override", IPHONE_USER_AGENT_V1_1);

    WebDriver webDriver = new FirefoxDriver(profile);
    return webDriver;
}
```

- We can also include extra Firefox extensions, e.g., Firebug when creating the FirefoxProfile, which may help when debugging our automated test.

---

[1]See the section on using an embedded browser for more details of the iPhone's webview component.

CHAPTER 6

Client Applications

Client applications are installed onto mobile wireless devices such as mobile phones. They are able to provide users with more functionality, better integration, and better performance than browser applications. Client applications may be written in a portable programming language (e.g., J2ME or FlashLite) or written as native code (e.g., Symbian C++ for Nokia phones).

Client applications can be divided into two categories: portable applications and native applications. Portable applications run in (or on) a virtual machine, such as the Java Virtual Machine. Native applications are written to run directly on particular architecture or platform of the target devices; for instance Nokia's Series 60 3rd edition platform.

Portable applications are written to run on an idealized platform, which should provide a consistent environment regardless of the actual details of the physical device. They are therefore less likely to have access to the latest and greatest features offered by a particular device. However, one application should run, essentially unchanged, on hundreds of different devices and models.

Native applications, in contrast, do have access to specific features provided by a physical device, and have access to things like the underlying file system, the contacts list, etc. They tend to run faster, but distinct, custom versions of an application need to be written to suit different target platforms. A target platform may be broad or narrow depending on the amount of integration with particular features provided by the devices. For instance, a target platform for Nokia phones could be "Series 60" that covers around 50 models, or subdivided into "Series 60 1st edition," "Series 60 2nd edition," and "Series 60 3rd edition" depending on whether the software needs to use specific features offered by later versions of the handset models.

6.1 PORTABLE APPLICATIONS

Sun was one of the first companies to try to provide a standard platform across many mobile devices where a program written and compiled for that platform can be installed on any of those devices. Sun's platform is known as Java Micro Edition (Java ME). Until recently Java ME was called Java 2 Micro Edition (J2ME), which is still the more common term for the platform and programming language. Subsequently, there have been several other "portable" platforms for mobile devices including:

- FlashLite (from Adobe),
- JavaFX (from Sun), and
- SilverLight (from Microsoft).

I will limit my discussion to Java ME as I have no experience of automated testing for the other platforms.

Although Java ME is intended to be a common, consistent platform, in practice many of the features were optional and manufacturers did not implement all of them, and some features were implemented partially. Many features are documented through Java Specification Requests (JSRs), e.g., JSR75 for access to the file system.

The software is bundled as a JAR file, together with a JAD file that describes the JAR file, e.g., the file size and the names of the main class. In practice a JAR file's internal structure is similar to a zip file and can be opened with the same software utilities.

Java ME is a reduced version of Java (standard edition). Java ME lacks some of the features of Java for development and debugging. For example, many devices offer only limited stack traces, and custom class-loaders are not available.

6.2 NATIVE APPLICATIONS

Native applications can take advantage of all the features offered by particular phone models, e.g., with very well integrated user interfaces, access to contacts, the camera, etc.

Native applications are often developed in C/C++ or a similar low-level programming language. Binary Runtime Environment for Wireless (BREW) is also considered a native platform, which is implemented on a relatively small number of devices.

6.2.1 Developing Native Applications

The compilers and development environments are more demanding than the equivalent tools for J2ME applications. For instance we install at least four heavyweight SDKs, an emulator, and Visual Studio in order to build an application for Windows Mobile phones.

Some of the software tools used to build native applications are easy to incorporate into automated builds and testing. However, others require significant effort just to run and require additional checking after the event to find out whether something went wrong or not.

6.2.2 Example Problems for Portable Applications

- One UK branded Sony Ericsson W880i does not stream audio (a generic W880i does). The application being developed had to be changed to wait for a defined period rather than waiting to receive a finite amount of data from the device.
- Access to the file system is provided through JSR75. A common J2ME application was written to access photos from a phone's file system. One manufacturer's phone displayed over 23 pairs (i.e., > 46 prompts), asking the user for permission, another asks 7 times, and a third manufacturer's handset thankfully simply crashed!
- Users complained that an email application did not load on various phones because it was too large. The information on the maximum JAR file size is inconsistent and unreliable so I ended up writing a test application where we could control the size of the JAR file in order to determine the maximum size of JAR file supported by various phone models.

6.2.3 Example Problems for Native Applications

- The expected API to allow SMS messages to be sent from the Windows Mobile version of Google Maps for Mobile was not available on one manufacturer's phone. The manufacturer provided details of an alternative API that worked once we had implemented it.
- Google Maps for Mobile on a new popular phone model had extraneous text in textboxes. The issue appeared to be related to the changes in the newer version of the phone's operating system, which we had to account for.

6.3 TESTING STRATEGY FOR CLIENT APPLICATIONS

Typically, testing client applications are done interactively, even when the tests are automated for various reasons, such as:

- Applications include a high percentage of UI code that needs to be verified visually.
- The software and development environments are more constrained and the tools more specialized to meet the needs of individual target platforms.
- Unit testing frameworks are less well developed for automated, headless (i.e. no user interface) testing.

In terms of an automation strategy for client applications, we combine automation with manual (human) checking, particularly for on-screen rendering.

Custom test applications are a good way of learning about the key characteristics of a device or its virtual machine environment (e.g., for Java ME). These can often run unattended and report results directly to a server for processing and analysis.

. . . .

CHAPTER 7

Testing Techniques for Client Applications

In my experience client applications are harder to automate. This is the area with the richest potential in terms of improving the quality and effectiveness of automated testing. Here are examples of techniques I have worked with.

- Automated unit tests: these are relatively well understood and commonplace for testing client applications. The unit tests often run within the relevant simulator, e.g., Sun's Wireless Tool Kit's (WTK) emulator for Java ME applications.
- Code instrumentation: particularly relevant for automating tests for Java ME applications.
- Custom test applications and prober applications: these are a good way to explore the capabilities and quirks of actual devices, e.g., to test how access to the file-system (using JSR75) behaves on each phone model.
- Signature testing: we can create UI-less clients that behave like the full client at the network interface. To the server they are essentially fully-operational clients; however, they only contain the essential code required to perform the network I/O.
- GUI-level automation: including optical character recognition (OCR) and encoding text in image pixels. These are both described in the common techniques chapter.
- Using "mock" libraries, e.g., Hammock for Java ME.

As part of our work on test automation we are working on making improvements in terms of test automation for these applications.

7.1 AUTOMATED UNIT TESTS

Unit test frameworks exist for most mobile development platforms. In some cases these are specific to the mobile device platform, e.g., Symbian OS unit that works for Series 60 and UIQ devices. In other cases there are a number of competing frameworks, e.g., for Java ME.

The tests are often executed in an emulator or simulator, although some development platforms allow the unit tests to be run on actual phones. Some unit tests (those without a UI component) can generally be fully automated (e.g., using J2MEUnit for J2ME applications), where such tests can be run using the Text-based test runner. However, in practice unit tests are often unable to test things like the GUI layer, event handlers, or multithreading (which is why we need other ways to automate the testing). A useful article on J2MEUnit and Eclipse is available online (http://efforts.embedded.ufcg.edu.br/javame/?p=11).

Often the test logic needs to be embedded with the application code being tested, and can significantly increase the size of the application. In some cases devices with limited resources may not be even able to install or run the application.

While these tests are useful for development, as they enable individual pieces of code to be tested, they seldom scale to system or acceptance tests.

7.1.1 Examples of Unit Tests

Here is a unit test, written in JUnit, for a method that encodes strings in Base64 (Base64 allows all characters to be encoded in a safe set of 64 characters that should not be interpreted by the communication channel, in this application it is used for simple user authentication).

```
protected void testBasicAuthForEmptyStrings() {
        String authResult = BasicAuth.encode("", "");
        assertEquals("EmptyStrings should result in 4 byte result", 4,
authResult.length());
}
```

7.1.2 Running Unit Tests in an Emulator

There are several challenges when running unit tests in an emulator, including capturing the output and automating exection of the tests in the emulator.

Capturing the output. Many emulators were designed to be run interactively, where displaying the test results on a screen is sufficient. There are several forms of output, including the rendered UI, sound, events (e.g., vibrating the "phone"), and debugging statements written to the standard output and error consoles.

Screen capture code helps to record the rendered UI. And mapping the virtual storage card to a specified directory on the machine being used to run the tests may enable your script to capture the output from the tests.

Automating execution of the tests. One benefit of having automated tests is the ability to run them often, and automatically. While most emulators can be started automatically, and include command-line parameters to specify the test program to run, getting an emulator to quit automatically after the tests have completed can be a challenge. Also, you may want the tests to run in the background or on a machine without a GUI interface.

If you need to fully automate unit tests in such emulators you may need to devise creative solutions like using virtual graphical terminals in order run without a GUI, screenshots to determine when the tests have completed, and remote key entry to quit the emulator after the tests have completed.

7.2 SYSTEM TESTING FOR iPHONE APPLICATIONS

Native applications have been available for the iPhone since the launch of v2.0 of the iPhone device software. These applications are written in Objective C. A unit testing framework is available on code.google.com (http://code.google.com/p/google-toolbox-for-mac/) that even supports basic UI testing (http://code.google.com/p/google-toolbox-for-mac/wiki/iPhoneUnitTesting).

At Google, we also found a way to include system-level tests for native applications. Future versions of this document may include more information. Until then please contact me directly if you need some tips on how to get started.

7.3 CODE INJECTION

Currently we have successfully used code injection for J2ME and Blackberry applications.

One option for test automation is to inject code to test and monitor an application. An existing application's binary software is processed to add additional object code. The injected code may simply be used to obtain information, e.g., statement coverage, timing data, etc., or it may include automated tests.

Code injection is significantly more involved than unit testing or incorporating system tests at the application level; however, it provides several unique features not available from other forms of test automation:

- Source code is not required for some forms of code injection, e.g., to add code coverage instrumentation.
- The application's source code does not need modifying.

Injecting code requires:

- The application,
- The code to inject (which may be tightly coupled to the implementation details of the application),
- A way to inject the code, and
- Instructions on where and how to inject the code.

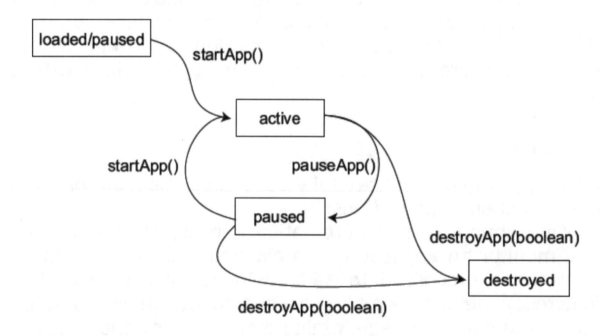

FIGURE 7.1: Java ME life cycle.

7.3.1 Code Injection for Java ME

This section provides an overview of the injection process. The details are more involved and creating effective support libraries (to interface tests to the application being tested) may take weeks or months of effort, although my colleagues and I are working on finding ways to simplify and reduce the effort required.

Java ME applications conform to the MIDlet model shown in Figure 7.1 (reproduced with permission from Stefan Haustein who coauthored MIDP Programming with J2ME, by Sams Publishing). At least one chapter, Chapter 3, is available online (http://www.developer.com/java/j2me/article.php/1561591) if you would like to learn more about J2ME development. A good understanding is important to perform the code injection effectively.

Java ME does not allow other software to start a MIDlet for security reasons, and it also does not support reflection or introspection—both available with Java Standard or Enterprise Editions. Also, the security model prohibits ClassLoaders to be overridden. Typically test automation tools for Java rely on techniques such as using custom ClassLoaders or reflection, so unfortunately none of these tools work for J2ME.

Java ME also places a number of practical constraints in terms of test automation. The software has to run with limited resources and under the constraints of Java ME's security model, which prohibits shared memory and limits the application to a sandbox. Java ME software may be permitted out of the sandbox, e.g., to access the filesystem through a combination of optional JSRs (e.g., JSR75 for filesystem access) and security permissions (e.g., a signed application may have some restrictions lifted). Implementation of JSRs is the responsibility of whoever provides the Java Runtime for a given device model, they choose whether to implement a JSR, how to implement it, and how much of the JSR to implement. The provider is also hard to pin down as it tends to be a combination of several parties including the manufacturer of the device and the network-provider. All in all, quite a challenge to address!

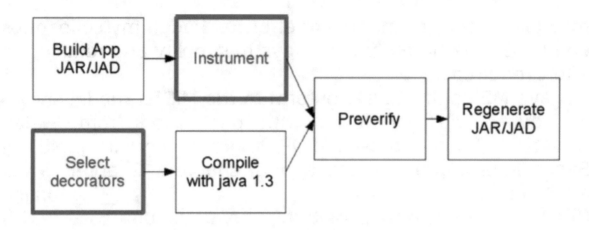

FIGURE 7.2: An overview of code injection for Java ME.

Code injection for Java ME consists of several related activities:

- Preparing the application so it can be instrumented effectively, e.g., by disabling optimization and obfuscation of the byte code, and by adding debug information to the compilation stage.
- Unpacking the JAR file if we are waiting until the end of the build process to inject code rather than doing so in parallel with building the application.
- Creating support, or helper, classes (e.g., to interface with a graphics library), to add code coverage (a related activity that also requires code injection), or to access internal data structures.
- Designing and implementing the actual tests (why we are doing all this work in the first place!).

After any code injection the resulting application needs to be processed (in the same way that any Java ME application has to be packaged for release), the resulting code needs to be pre-verified, and a (new) JAD file created. Figure 7.2 provides a pictorial overview of the process.

The steps can be straightforward, but rely on a good understanding of the object code or byte code. Often a tool or software library may be available to simplify the instrumentation process.

For Java ME applications, a number of mature open-source libraries are available to instrument the code, e.g., ASM (http://asm.objectweb.org/index.html).

Implementing the tests. One effective way to implement the tests is to have them run in a separate thread within the application being tested. The tests need to be started by adding a hook into the startApp() method, and hooks should also be added to pauseApp() and any destructors (to clean up generally and store the test results). The tests need a way to determine the state of the application (e.g., so they can confirm the application is on the home screen), some way to trigger events (e.g., by inserting a key press representing the "select key"), and ways to detect whether the expected behavior occurs (e.g., that the menu is displayed with the correct items, in order, in the appropriate place on the screen). We created support libraries in Java to perform all of these steps.

Miscellaneous comments. The techniques work across a broad range of platforms, including Blackberry. However, some tweaks may be required to enable the automation to work for each platform.

Several colleagues at Google (Michele Sama and Olivier Gaillard) and I presented a talk at GTAC in October 2008 on JInjector, a tool we developed in 2008 that significantly simplifies the effort required to use code injection for J2ME applications. A video of the presentation is available at http://www.youtube.com/watch?v=B2v5jQ9NLVg, and the code is available online at http://code.google.com/p/jinjector/.

7.3.2 Custom Test Applications

Custom "test" applications can be developed to test key behaviors, e.g., whether a given network supports UDP traffic, whether a Java ME application has access to the file system, the performance characteristics of the record store on the phone, etc. These may be developed in the same programming language and development environment as the main application (e.g., in Java ME) or in a different language (e.g., Python), depending on the skills and needs of the creator.

7.4 PROBER CLIENTS

Prober clients are similar to custom test applications; however, they are generally written in the same language and programming language as the main application, as they need to run on the devices. They can also help us to learn more about the behavior and capabilities of client platforms quickly and without needing fancy UIs, etc. They are particularly useful for Java ME applications. Examples of the types of tests that can be implemented include:

- JSR support (e.g., JSR75 for the file system and JSR135 for multimedia).
- RMS performance characteristics (data storage for Java ME applications).
- Memory management.
- Other performance tuning.
- Size of run-time for the application.

We can also test whether signing the application helps reduce the number of interactive prompts users have to acknowledge (Java ME applications).

7.5 SIGNATURE TESTING

Debugging issues when a client application is installed on physical devices can be very tough. For instance we were testing an application that uses UDP to send content to the servers. The testing was being performed remotely by manual testers who had problems getting the application to work end-to-end; and the application still needed polish so the error messages were not ideal.

We needed a way to identify, isolate, and address the connectivity issues. We created a UI-less client which incorporated the essential network characteristics of the full client. Our client ran on a desktop machine, and connected using a GPRS modem to the mobile network, and then to our servers. Our client could exercise the items identified with a tick in the following list of factors that could prevent end-to-end communications from working:

- ✓ Server application error(s)

- ✓ Server configuration issues

- ✓ Firewall and related components

- ✓ Significant packet loss on networks or interfaces

- ✓ Carrier network issues/blocking

- ✓ Features supported by customer's mobile contract, e.g., WAP only?

Other factors that can prevent end-to-end communications include:

- Firmware/network operator customization on the device.
- Client application bugs/configuration.

- Incorrect configuration, or selection, of the network settings (known as the Access Point Name [APN]) on each device.

Our client had the same "signature" in terms of the network behavior and helped us to test the APN and carrier connectivity independently of requiring manual testing using the full client.

APN consists of a set of parameters to enable network connectivity for a given network operator and data plan. At the time some of these network settings were hard to obtain and the available information was not always accurate.

On some devices users were prompted for the APN when installing the software. Other devices asked for them when starting the application. And in some cases the devices "remembered" the setting, which was a problem if users had not picked the correct APN at the time, especially as sometimes there was no way to correct the mistake afterward!

7.6 TEST TOOLS FOR CLIENT APPLICATIONS

There are at least three ways automation can help with testing client applications, including on- device debugging, test automation for the test environment, and automated unit tests, mentioned earlier in this chapter.

7.6.1 On-Device Debugging

On-device debugging allows developers to control and inspect their software on real devices. Although on-device debugging is not really a way to automate the testing, it can help us to find out what is going wrong.

Some work has been done by commercial software vendors to debug software on physical phones. Companies such as SonyEricsson provide tools to enable software to be debugged on their more powerful phones. Version 3.0 of Sun's Software Development Kit (SDK) for Java ME supports on-device debugging for phones that use ActiveSync; and Microsoft provides a highly-integrated development environment for their C++ software platform that allows software to be debugged on devices that run the Windows Mobile operating system. Android supports on-device debugging; and as the platform is freely available as open-source code, there is plenty of scope to create and customize the debugging tools.

7.6.2 Test Automation of the Runtime Environment

When we are able to automate the run time environment we can control and interrogate a running application. Ideally the test automation would include interaction with the native system events (e.g., for keyboard and other inputs) and the ability to query the GUI layer (e.g., to read the contents of a text box). Extra features could be provided to enhance our ability to test the application, e.g., where system calls can be intercepted and modified to force certain conditions to be triggered. In comparison, tools such as Security Innovation's Holodeck provide these capabilities for the desktop and server versions of Microsoft Windows operating systems.

I am not aware of any companies providing public support for automated testing on their devices. Instead, I know of several ad-hoc solutions that enable tests to be automatically deployed and executed on devices.

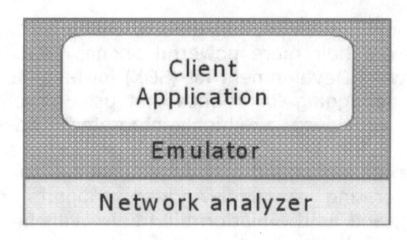

FIGURE 7.3: Monitoring network traffic when running a client application in an emulator.

There are a few examples where emulators include an automated test framework (ATF), for instance Sun demonstrated an ATF for version 2.6 of the Java ME WTK at the JavaOne Conference in 2008. However, I do not think any of the automation frameworks have been made generally available.

Typical challenges when automating tests using emulators are:

- Driving the user-interface.
- Interpreting screen-responses, including stitching together images for pages that require scrolling, pattern matching where the content varies from request to requests.
- Latency in the device interaction.
- A lack of fidelity between the behavior of the emulator compared to actual devices. There may be serious issues with your software, even if your tests pass on the emulator (Figure 7.3).

7.6.3 Emulators

Emulators are ubiquitous for client applications, and are often provided by the manufacturer or software provider in the case of Java ME (see screenshot).

As few emulators provide support for automation. Automated testing tends to consist of building an application where the tests are incorporated into the application and run in an emulator. The tests (or the test framework such as J2MEUnit) report the results, which are then checked for errors and other issues.

Characteristics of using emulators. Emulators run on a desktop computer and may be limited to a single operating system (generally Microsoft Windows). They are easy to use interactively and often faster, and slightly easier to use, than using physical devices. The performance characteristics are very different from the actual performance of physical devices, and in our experience we have to test performance optimizations on devices as we got misleading results when running these tests in an emulator (Figure 7.4).

FIGURE 7.4: An emulator.

Some, such as Sun's Java ME emulator, emulate general characteristics (e.g., color screen and a QWERTY keyboard). Other emulators include relatively faithful representations of specific phones (e.g., the BlackBerry development kits). However, they seldom emulate the quirks and bugs of actual devices, which makes testing some bugs and resulting workarounds very difficult.

Emulators might run on a different platform (Windows on an x86 processor) than the target platform (mobile operating system (OS) on ARM). The differences may invalidate tests that cross the application boundary, such as:

- For performance and benchmarking: since the code is running on a different processor, it is difficult to estimate and instrument both CPU and memory usage; and

- OS/API assumptions: since the code is running within a host OS where the emulator translates calls to the underlying OS or runtime platform. The translation makes it very hard to detect device-specific bugs.

As mentioned earlier, emulators are seldom designed to be automated, and so require significant effort to implement even a flawed automation. They tend to be proprietary and therefore difficult to integrate with automated testing. Logs and data files are often hard to access, although some vendors may provide facilities to copy files and data to and from the device.

Examples of emulators. Each handset manufacturer and/or operating system provider offers their own, distinct approach to using emulators.

Examples of emulators for native applications:

- Microsoft's emulator emulates the ARM processor used in Windows Mobile devices. It emulates the operating system and the application running in the emulator could also run directly on the phone.
- Nokia's Symbian and SonyEricsson's (UIQ) emulators do not emulate the underlying hardware and developers need to build their applications for either the emulator or for the physical devices.
- PalmOS applications are built as 68K code (i.e. as if they would be running on the Motorola 68000 CPU series). Some speed-critical subroutines can be written as native ARM subroutines. However, the emulator does not run the ARM subroutines, instead developers need to build their subroutines as a Windows DLL or compile them for 68K to run on the emulator.

Examples of emulators for J2ME:

- Sun's WTK, which runs applications in a Java Virtual Machine. The resulting code can also run unchanged on a phone.
- There are a number of other emulators provided by handset manufacturers, including Motorola and Nokia.

An open-source emulator, called MicroEmulator, exists for J2ME. It provides a "glue" layer to execute J2ME code in a J2SE environment. Other people have enhanced the source code to implement additional JSRs. Note: As MicroEmulator actually uses J2SE libraries rather than J2ME libraries (which are much more limited and have some behavioral differences) some classes of bugs are not detected using this emulator.

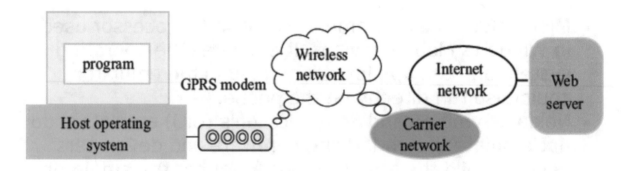

FIGURE 7.5: Connecting a program in an emulator over a PPP connection. OTA.

Using emulators OTA. When an emulator is able to connect using the generic networking capabilities of an operating system then we can configure the test environment so that the network traffic goes OTA, which provides more realistic performance characteristics than using a local area network (LAN). The data connectivity section of this material describes how to connect a computer OTA.

After configuring and testing the OTA connection, we need to configure the computer so little or, ideally, no other network traffic uses the OTA connection, otherwise that network traffic may adversely affect the results (Figure 7.5). To reduce the traffic, try to suspend background updates (e.g., by antivirus software, quit web browsers, shutdown all but the essential services and utilities, etc.).

7.7 TEST AUTOMATION USING THE BLACKBERRY EMULATOR

The BlackBerry device simulator from Research In Motion (RIM) supports automation. Command-line tools are used to enter commands, and to create screenshots (Figure 7.6).

Key elements include:

- Fledge.exe—the simulator, which can be started from the GUI or from the command line
- Fledgecontroller.exe—the input mechanism, allows various inputs to be sent to the simulator, e.g., key presses
- Javaloader.exe—allows screenshots to be captured and saved as bitmap images. Javaloader uses a USB connection to grab the screenshot from a device or emulator. Therefore the USB connection needs to be enabled for the emulator. The USB can be enabled in the simulator's UI, using the Simulate > "USB Cable Connected" menu option, or with a command line option: /execute=UsbCableInsertion(true)

FIGURE 7.6: Test automation using the BlackBerry emulator.

Examples of using the tools include:

```
simulator_path\simulator\fledgecontroller.exe /session=8800 /
execute=KeyPress(a)
```

```
simulator_path\bin\JavaLoader.exe -u screenshot c:\screenshot.png
```

7.7.1 Summary of Testing Techniques for Client Applications

We have a wide range of test automation techniques available, ranging from unit test frameworks to hardcore code injection. Custom clients and prober applications can help discover issues with devices or the environment quickly and effectively, rather than trying to debug these issues through the application code. While emulators are useful, they are seldom intended as test automation tools.

CHAPTER 8

Common Techniques

Some automation techniques apply for multiple classes of application. These are described in this section.

- Image stitching is used to reconstruct an image of information rendered on a physical device or on an emulator.
- OCR matching.
- Encoding data in pixels.
- Model-Based Testing (MBT).

8.1 GUI-LEVEL AUTOMATION

Software agents enable GUI-level automation. However, challenges (in terms of making the tests relatively reliable) include:

- The need for high contrast, opaque, block colors for backgrounds, etc., to make images easier to match for menus, etc.
- Deciding whether to use image pattern-matching or OCR.
- Structuring test resources, e.g., you may need a set of images per device.

8.2 IMAGE STITCHING

Image stitching is used to assemble a number of screenshots into a composite single image that represents the original page. It is particularly useful when the page is too large to be displayed on a physical phone. There are four steps involved in image stitching:

8.2.1 Steps in Image Stitching

Step 1: Image retrieval. This is fairly straightforward. Assemble a set of screenshots of a web page by scrolling down at the finest granularity the device allows (or, alternately, the granularity level determined through heuristics and statistical analysis). The way and amount may vary between different models or even versions of the installed software. And the scrolling may be affected by the navigable elements in the UI (e.g., web page links) and input fields on an electronic form.

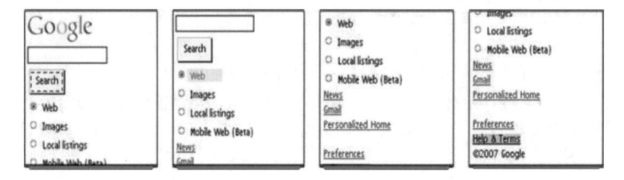

FIGURE 8.1: A set of example screenshots.

The following screenshots represent a simple example—the real one may involve as many as 30 screenshots to travel the same distance (Figure 8.1). We can assume at least one screenshot for every navigable element that appears on the screen.

These four images represent the four images we will use to reconstruct the full web page. As you can see, there are slight variations in the images: in the first image, the search button is selected; in the second, we have moved the selection down one to the web link.

Step 2: Image alignment. Image alignment is the process of finding the correct offset of the new image from the image that came before. An example of the alignment is provided in Figure 8.2.

FIGURE 8.2: Aligning the screenshots.

The alignment can be performed by brute force, trying to compare the pixels row by row between the two images. The best match is the first offset with the lowest number of differing pixels. A maximum offset is specified, and should be no more than the height of the phone's screen. Other parameters are used to reduce the chance of false matches.

Step 3: Image subdivision and election. Once we have our images in their final, absolute positions, we can figure out which parts of which images we wish to stitch together into a final representation of the page.

This is done through a kind of voting process—we perform a popularity contest for each part of each image represented within the complete page. To do this we first figure out what the image boundaries are—a divider is placed for the start or end of each image. Once all the dividers are in place all of the images are tested to see whether or not they fall within a particular region; if they do, that region is cut out of the original image, and held in an object, in memory, which tracks how often a particular image is seen.

As you can see in Figure 8.3, the second row has a search button. The first is selected; the second is not. In this example we have two distinct candidates and two votes, one for each. Both are equally popular. In this case, we choose the one to appear last.

The third "row" is the more general case—the same section of image will appear multiple times as the scrolling was recorded; one of those scroll events will represent a link or form element in its selected state. In these cases we have multiple possible images to choose from; two, however, are identical, so there are two distinct candidates: the middle one with one vote (the one with "web" in blue) and one other represents both the left-hand and right-hand slices of this "row," with two votes. The image with two votes (i.e., without the text selection) will be elected as the best representation.

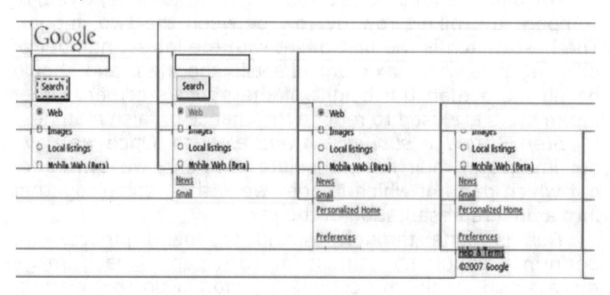

FIGURE 8.3: Slicing the images into rows.

Step 4: Image reconstruction.

Once we picked the representative images for each "row," we can stitch them together, one after the other, to reconstruct the final image.

While this is a trivial example, this can be scaled up and out in multiple ways—reversing direction, supporting horizontal as well as vertical scroll, etc.

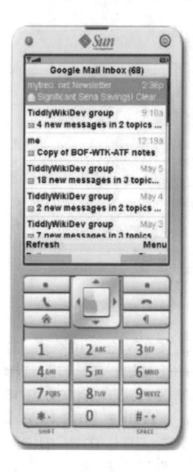

FIGURE 8.4: How the content varies across screen sizes and resolutions.

8.2.2 Challenges of Image Matching

In theory, matching images is relatively straightforward, especially once transparency is used. However, in practice the content validation may require lots of work to implement as the phone platforms are so diverse and the conditions change frequently, possibly every time the tests are executed. As an example, if we decide to automate the testing of an email client the captured images may vary significantly from device model to model. Here are some screenshots of an email inbox for three variations of phone (Figure 8.4).

8.2.3 optical Character Recognition

OCR has been used for years to extract text from images. OCR applications are even provided with entry-level document scanners. Conceptually, they extract textual content from images and return the content as text which can then be used and edited more easily than trying to do so with the image. For mobile test automation OCR offers the potential to extract text from GUI screenshots, amongst other things.

For automated tests OCR offers the potential to reduce the need to compare images, rather we could compare the text, e.g., of a menu to determine whether the correct values are present.

However, based on my experiences of using several open-source and commercial OCR software libraries the extracted text is both incomplete and wrong. The error rate seems around 40%. Text on a mobile phone screen tends to be quite small (e.g., 8 point), possibly too small for general purpose OCR libraries.

Currently, I cannot recommend using OCR in automated tests for mobile wireless applications. However it may be possible to train an OCR library to improve the accuracy of the recognition. I would like to use OCR once it is sufficiently accurate and trustworthy as it offers the potential to significantly increase the reliability of the automated tests while reducing the need for sets of screenshots per device.

8.2.4 Encoding Data in Pixels

Most things on computers are represented in memory, including text and images. Images are encoded as data, which is interpreted by programs and devices when rendered. Text is also stored as data.

Each character of ASCII text can be stored in 8 bits of data. A common notation of 8 bit data is a hexadecimal code in the form 0xnn where *nn* represent two hexadecimal characters in the range 0-9, A—F. The ASCII character codes for common characters are consistent and well documented, e.g., the letter capital "A" is represented as 0x41 in hexadecimal, while lowercase 'a' is represented as 0x61.

Each digit of hexadecimal represents a "nibble" of computer memory—4 bits of data.

HEX	0	1	2	3	4	5	6	7
Binary coding	0000	0001	0010	0011	0100	0101	0110	0111

HEX	8	9	A	B	C	D	E	F
Binary coding	1000	1001	1010	1011	1100	1101	1110	1111

Conceptually the letter "A," therefore, is represented in memory with 01000001; and the letter "a" with 01100001. The essential difference between lowercase and uppercase letters is that the 3rd bit from the left is set to 1 for lowercase letters and 0 for uppercase letters.

Based in this encoding, the text "Help" would be represented as follows:

First in hexadecimal characters: 0x48 0x65 0x6C 0x70
And in binary: 01001000 01100101 01101100 0111000

(The underlying computer architecture may actually store the information in a different order, which is one thing we need to check for later on in this test technique).

Now let us consider how images are stored on a computer or phone. Typically, each pixel (the smallest dot on the screen) is encoded using 32 bits as follows:

VALUE	TRANSPARENCY	RED	GREEN	BLUE
Range	0 to 0xFF	0 to 0xFF	0 to 0xFF	0 to 0xFF

Effectively, a pair of hexadecimal characters represents each value. A solid white pixel would be represented as:

0xFF 0xFF 0xFF 0xFF in hexadecimal
or 11111111 11111111 11111111 11111111 in binary

Here is where we can take advantage of how pixels are encoded to encode text within pixels. For instance the string "Help" could be encoded in a single pixel! The color would be a mucky gray (since the character values are relatively similar). However, if we are able to detect the "hidden" characters (e.g., by agreeing which pixel(s) they will be encoded in), then our automated tests can easily and reliably extract the text, rather than relying on image matching or OCR.

We can build on this technique to pass the text contents of menus and other text displayed on screen and even other data such as status codes, all within a few pixels on screen.

The following figure demonstrates the concepts of providing the text "Julian" within a few pixels at the top left of a screen. In this case, I have used artistic license by using a few green pixels to brighten up an otherwise dull series of gray pixels.

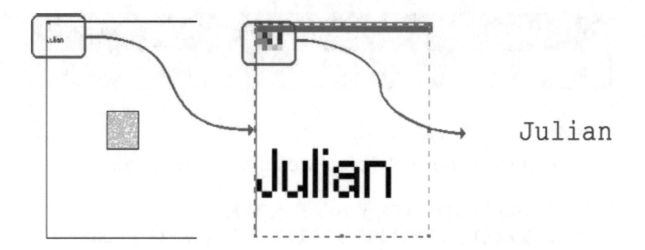

Tips and traps. As I have already mentioned, the computer may actually store text and/or images slightly differently based on the underlying hardware, for instance. Your code may need to change the order of data, either when it is written as pixels or when the pixels are extracted from the image of the screenshot.

Some devices may support different "color-depths"—the number of bits used to represent each color, e.g., modern desktop computers may use between 16 and 64 bits to represent each color. Also, characters may be represented in a variety of formats including various Unicode formats. There are various ways to address these issues, e.g.:

- Write a signature sequence at the start of the set of encoded pixels. When the signature is decoded, the code used to read the signature can pick the appropriate formula, or pattern, to use when matching the main encoded data.
- Repeat the data several times within each pixel and discard duplicates when reading the image.

Some devices may use color pallets (a sort of lookup table) to match colors to pixels. These lookup tables may limit the number of characters we can safely and reliably encode in a pixel. Furthermore, intermediate programs such as VNC Viewers and the GUI interface of the computer receiving the image may all reduce the fidelity of the encoded data. I suggest you create some tests to determine whether this is an issue for you in practice.

Web-safe GIF images are an example of how a limited pallet is used to represent colors across a broad range of web browsers.

In summary:

- We can incorporate text data in image data,
- The text is rendered as seemingly "random" pixels,
- The text can be reconstructed from image data,
- Test code can process and match reconstructed text, and
- The encoding may need to include error-detection and error-correction to compensate for palette tables, both on the device, and in each intermediate dynamic image.

8.2.5 Making Image Matching Easier

Here are a number of tips and techniques that may help to make image matching easier for your automated tests.

- Automate screenshots so a distinct screenshot is captured after each controlled input (some inputs such as network errors may be outside your direct control). For example, after every key event a script could take and store a screenshot, and store it with a monotonically incrementing filename (the first file is called image1, the second image2, etc).

- Find, adopt, or even create good, efficient image processing tools to reduce the burden of editing, cropping, and otherwise manipulating images.
- Store the images for each distinct model of device in a separate file directory. The name of the directory can include the name of the device, e.g., /nokia6230i. If you need to capture separate images for sub-models or custom firmware (e.g., when the device is provided by a particular network operator), then include enough information in the directory name to identify the specific nature of the device.

Autogeneration of images. With a small helper application you may be able to autogenerate images for known text on a given device. Conceptually the application would "echo" each input character on screen. The screenshot can be captured automatically in the controlling program and processed to extract the image that represents the characters to match when testing the application we need to test. The controlling program would need to specify the font-face (e.g., Arial, Courier New), the style (e.g., bold, italic), and the font size (e.g., 8 point) to enable the correct image to be generated.

8.2.6 Using Advanced Image Matching Techniques

Image matching is an active field of research and practice, e.g., for automatically matching faces on photographs, matching iris scans for biometric security, etc. I suspect some of the concepts used for image matching in general may help to improve the efficacy and reliability of matching screenshots from mobile wireless applications.

8.2.7 Detecting Good and Bad Results

Another challenge is to recognize whether the results are good, or problematic, purely from the screenshots. The following figure includes two screenshots from a mapping application. Can we tell from either screenshot in Figure 8.5 whether something is wrong?

Screen A

Screen B

FIGURE 8.5: Is there a problem in either of these images?

For these two screenshots, Screen A is displayed while waiting for content to be downloaded and rendered. Here we may want our script to wait before capturing another screenshot, by which time the content may have been loaded correctly. Screen B includes red X's that are only displayed when a problem has occurred.

8.3 CONTACT SHEETS

One technique which has generated good results is to combine automated "user" input with manual verification of the resulting on-screen content. Events such as key-presses are automatically generated, and after a certain period screenshots are taken and stored for later comparison. The screenshots can be laid out as "contact sheets." Contact sheets provide a multicolumn format of rows of small images that can be used to quickly spot possible anomalies ready for more detailed investigation.

8.3.1 Using Transparency Masking

In order to increase the reliability of the tests, intermediate screens are matched with reference screenshots. The reference screenshots generally contain areas that do not need to be matched (for instance, where the content changes frequently and is not germane to our tests).

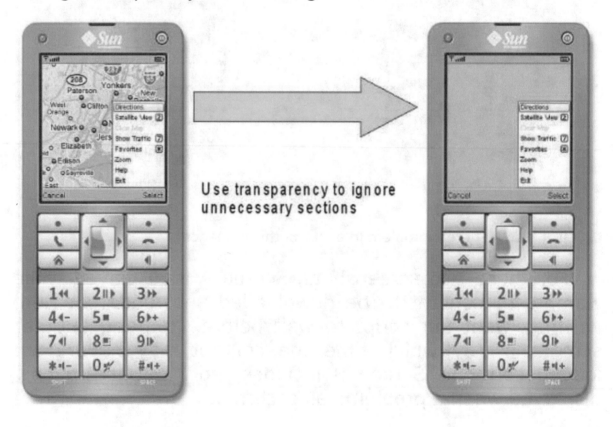

FIGURE 8.6: Using transparency to ignore unnecessary sections.

Original

Transparency mask in
black

FIGURE 8.7: Transparency masking of an image.

An example of something we often want to ignore is the time of day displayed on some phone screens. The term for marking the areas to ignore is called *transparency masking* (Figure 8.6). See Figure 8.7 for a simple example of a BlackBerry home screen.

Here we mask: the time, the date, all but 2 bars of the signal strength indicator, the connection type, and most of the battery strength. We want to fail the match if the signal strength is very low or the battery is fully discharged.

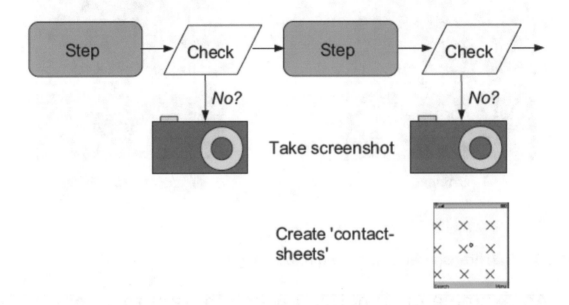

Best of both worlds
human-assisted automation...

FIGURE 8.8: Assisting test automation with human decision making.

By matching the screens it is also easier to reset the device to a known state (e.g., back to the idle screen) at the end of various tests.

8.3.2 Combining Automation With Human Judgment

Until we can reliably automate all aspects of our application tests, we can take advantage of the strengths of automated test scripts, which run lots of tests automatically and unattended, with the pattern matching skills of humans who can typically interpret images quickly and effectively (Figure 8.8).

8.4 MODEL BASED TESTING

Model Based Testing (MBT) is already used to test other software domains such as web sites, desktop applications, etc. It helps to automate longer-running automated tests, where the test software interacts with the software being tested and compares the actual behavior of the software being tested with the expected behavior contained in a programmed model of the desired behavior.

We have used MBT techniques to test a number of mobile wireless applications, including a search engine and client applications.

When testing a search engine, the test engine can be implemented in a very similar fashion to implementing model-based tests for web applications. The key differences involve:

- Manipulating and checking the HTTP headers in the requests and responses, and
- Dealing with a number of markup languages.

When testing client applications we decided to implement the model as a separate program that ran on a distinct computer, rather than on the client because the client application is already constrained by resource limitations (Figure 8.9).

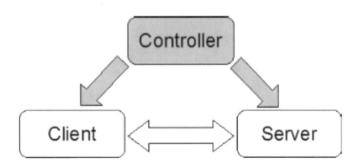

FIGURE 8.9: Using a separate controller for MBT.

The executable model interrogated the server for appropriate state, context, and data and interacted with the client application through a small software module that was incorporated into the client application. The executable model and the client module communicated OTA using proprietary messages, created specifically for the application being tested. The data structures used common formats used elsewhere on the project.

CHAPTER 9

When to Test Manually

During my many years of working with software I have heard and read numerous arguments for the benefits of automated testing. I agree with the goals and ambitions of automating as much testing as practical. However, I believe good software testers can often generate better results with manual testing, particularly when we use a hybrid approach that builds on the relative strengths of humans and automated tests. Also, I have lived through numerous projects where the automated testing lagged behind the software we wanted to test, and where—sadly—the automated tests ended up being a waste of time and energy, not having exposed a single issue.

Before I provide some examples where manual testing makes sense, here are some of my maxims for measuring the effectiveness of testing:

- What do we learn from the results of the testing? I am looking for things like bugs found, bugs prevented, increased and justified confidence in the system under test, etc.
- How much time was invested in creating and running the tests? The time should be divided into several chunks: the time required to run the first set of useful tests, and the time required for each subsequent deliverable, e.g., for each new release of the software being tested, or for each new release of automated tests.
- What problems did our tests miss? What questions remain unanswered?

9.1 EXAMPLES OF EFFECTIVE MANUAL TESTING

- UI and rendering problems can often be noticed immediately by a person. Common problems we can spot easily are alignment issues, flickering, etc.
- When the UI is changing frequently, test scripts tend to break and need repairing, requiring lots of work. However, if the development team is supportive of automated testing, they can help reduce the effort required to maintain and update automated tests, e.g., by using consistent labels for key elements.
- When content is dynamic and hard to predict. I wrote an article about using weighting to improve the accuracy of (automated) tests, which may provide some suggestions on how to automate tests in these circumstances (see http://www.stickyminds.com/s.asp?F=S11983_COL_2).
- When the cost of automating is likely to significantly exceed the costs of manual testing. However I would encourage you to consider scripting languages such as Python or Ruby and create some simple automated tests, e.g., to reproduce existing reported issues, which could be used to speed up regression testing.

9.2 COMPUTER-ASSISTED TESTING FOR MOBILE WIRELESS APPLICATIONS

I have had excellent results when I have found ways to augment my testing with automation. Here are a few examples to get you started:

- Automating setup and other preparation (see Jonathan Kohl's Man and Machine article for some useful ideas and tips, http://www.stickyminds.com/s.asp?F=S13122_MAGAZINE_2);

- Message generation, and test bots, which reduce the workload on the tester; and
- Contact sheets (as mentioned in the chapter on Common Techniques).

9.3 TESTABILITY FOR AUTOMATED TESTING

The design and implementation of the mobile wireless application can have significant effect on how easy the testing is to automate. In the worst case, automation may be impractical, e.g., if the data structures and names change dynamically, have cryptic identifiers, and where the software is a poorly defined blob of code.

When applications are designed to make automated testing easy, the automation code is simpler, faster to implement, and more likely to be accurate and robust.

The following article provides an interesting view on how to implement an API to increase the testability and is focused on Windows Mobile applications at the time of writing:

http://www.jamosolutions.com/documents/Automation%20deployment%20-%20best%20practise.pdf!

9.4 HOW TO IMPROVE TESTABILITY

Here are some tips on how to improve testability for each class of mobile wireless application.

9.4.1 Browser-Based Applications

- Add ID tags to the main elements the automated tests will interact with. Make the names understandable and keep them consistent across releases.

9.4.2 Client Applications

- Consider using high-contrast colors, visual markers, and even pixel-encoding to make GUI-based automation more reliable.

9.4.3 SMS Applications

- Provide a scriptable library or interface (e.g., available from Python) to make tests easier to create.

9.4.4 General

- Provide a complete API to enable the test automation code to test "below" (without) the GUI.
- In programming languages such as Java consider making methods and data protected.
- Consider using an xUnit structure for the test cases. The development team is likely to already use a unit-testing framework such as JUnit for Java, PyUnit for Python, etc. If the larger tests use the same unit-test layout then the developers are more likely to write and maintain the automated system tests, etc.

CHAPTER 10

Future Work

This book covers some of the basic approaches to test automation for mobile wireless devices. As mentioned in the introduction, some additional topics are available online in draft form at http://sites.google.com/site/mobilewirelesstestautomation/draft-material-on-mobile-wireless. At the time of writing. these include:

- SMS applications, including testing techniques,
- How to test applications that use WiFi connections,
- Common tools,
- Test automation for Android applications,
- Measuring end-to-end performance, and
- Running more tests on the devices.

There is a lot more to do! Your contributions and involvement can help us collectively to improve test automation for mobile wireless applications.

What would you like to do next?

To be continued…

Links and References

I hope you will find these references useful. I appreciate your comments and recommendations too.

Here are some online resources related to this book:

Blog: http://mobilewirelesstestautomation.blogspot.com/

Site: http://sites.google.com/site/mobilewirelesstestautomation/

A.1 TESTING MARKUP (WEB SITES)

http://ready.mobi/launch.jsp?locale=en_EN—Online checker that allows a web site to be checked in terms of readiness for mobile devices. Simple, but quite useful. There are other similar sites and services available; however this one has more of a testing focus. It implements the w3c mobileOK basic tests (http://www.w3.org/TR/mobileOK-basic10-tests/). The w3c have recently launched a beta version of an online checker http://validator.w3.org/mobile/ and http://validator.w3.org/mobile/alpha.

http://www.cameronmoll.com/mobile and http://www.cameronmoll.com/mobile/mkp/—Include test pages to try with your xHTML phone browser.

http://dev.mobi/node/472—Sample code a web site might use to detect whether a request is from a mobile device or not.

http://www.w3.org/2005/MWI/BPWG/—Links to an open-source implementation of the mobileOK basic tests (written in Java).

http://www.tagjam.com/headers.php—Displays the HTTP
    headers, good for obtaining headers such as user-agent
    for our tests.

http://www.developershome.com/wap/detection/—A useful
    introduction to the nitty-gritty details of HTTP headers
    for WAP and xHTML detection.

http://www.zytrax.com/tech/web/mobile_ids.html—A
    discussion about user-agent strings for various mobile
    phones, with lots of examples.

http://www.ericgiguere.com/articles/masquerading-your-
    browser.html—A very readable article on user-agent
    strings, how to change them in various desktop
    browsers, etc. The site includes various ways to view
    the user-agent sent by your web browser.

http://www.pctools.com/guides/registry/detail/799/—Details
    of how to customize the user-agent string sent by
    Microsoft's Internet Explorer.

http://uche.ogbuji.net/tech/4suite/amara/—The Amara
    Python module, used to parse xml such as xHTML and
    WML responses from web sites.

http://www.w3.org/Protocols/HTTP/HTRQ_Headers.html—Last
    updated in 1994(!) documents common HTTP request
    headers including User-Agent and Accept.

http://tuxmobil.org/phones_linux_wap.html—Summary of
    software suitable for WAP/WML and iMode/cHTML. The
    site has links on how to setup Linux with various mobile
    wireless devices, however a number of links are broken.
    ☹

http://www.diveintopython.org/http_web_services/user_agen
    t.html—Part of a great free resource on Python, this
    section describes how to set the user-agent string for
    HTTP requests.

http://www.crummy.com/software/BeautifulSoup/#Download
    —A very useful Python library to prettify web pages.

http://www.zvon.org/xxl/XPathTutorial/General/examples.html

A.2 J2ME TESTING

http://code.google.com/p/jinjector/—JInjector project homepage.

http://j2meunit.sourceforge.net/—J2ME unit itself.

http://www.wikistudent.ws/hammingweight/modules/hammock/—Homepage for Java ME mock objects representing common libraries. The source code is avialable at: http://hammockmocks.sourceforge.net/

http://kobjects.sourceforge.net/me4se/—Open-source project that provides the Java ME APIs in J2SE. See also http://midpath.thenesis.org/bin/view/Main/ for an implementation of MIDP that works with J2SE.

http://blog.emptyway.com/2007/04/05/comparison-of-java-me-unit-testing-frameworks/—Interesting comparison + good comments from readers.

http://java.sun.com/products/j2mewtoolkit/—Wireless ToolKit download link. Sun's toolkit for J2ME development that includes their emulator.

http://www.microemu.org/unittests.html—How to use the open-source MicroEmulator to automatically run unit tests for J2ME client applications. A worked example would be useful.

http://code.google.com/p/testingemulator/—Is an enhancement to the MicroEmulator and implements support for additional JSRs.

http://pyx4me.com/snapshot/pyx4me/pyx4me-cldcunit/—CLDC Unit test framework, untried by me.

http://developers.sun.com/mobility/midp/articles/test/—Fair article on using J2SE, lacks technical depth but written with a testing mindset.

http://www.devx.com/wireless/Article/32540—Worked examples of using J2MEUnit and JMEUnit testing frameworks.

http://weblogs.java.net/blog/alexeyp/—An interesting blog by a Sun insider who writes about ways to improve the testing of J2ME software.

http://developers.sun.com/mobility/allsoftware/—Sun's list of tools and emulators from various third party suppliers.

BlackBerry have several online articles, including:

- How To—Automate testing with the BlackBerry Simulator Article Number: DB-00531.
- How To—Use Javaloader to take a screen shot Article Number: DB-00484

I do not have a persistent link for them but search engines find them.

http://www.robotme.org/files/AutomatedGUITestingOfMobileJavaApplications.pdf—Masters thesis (also available in a similar form from Springer-Verlag to purchase) on how to automate the integrations testing of J2ME applications.

http://cobertura4j2me.org/—Open source code coverage tool for Java ME software.

http://www.sic-software.com/artikel/schulten_testen.pdf—An article, in German, on J2ME test automation.

A.3 JAVA BYTE CODE INSTRUMENTATION

http://www.ibm.com/developerworks/ibm/library/it-haggar_bytecode/—A useful introduction to Java bytecode.

http://en.wikipedia.org/wiki/Java_byte_code—A brief example of Java bytecode, including a summary of each bytecode.

http://asm.objectweb.org/index.html—ASM bytecode framework, useful for implementing bytecode instrumentation.

A.4 NATIVE APPLICATION TESTING

http://www.symbianosunit.co.uk/—Unit testing framework for Symbian phones.

A.5 TEST AUTOMATION WITH EMULATORS

http://www.perftestplus.com/resources/EA.pdf—A useful article on test automation using emulators. While the tools and emulators are old (circa 2001) the concepts and examples are relevant. Here is a related set of slides: http://www.perftestplus.com/resources/EA_ppt.pdf

A.6 SMS SERVICES

http://linux.softpedia.com/get/Communications/Telephony/SMS-Server-Tools-5735.shtml—Open source SMS tool, uses a GPRS modem to send and receive the messages.

A.7 CONNECTIVITY

How to connect your computer to the Mobile Wireless network(s).
http://umtsmon.sourceforge.net/—A very useful Linux package that configures and connects to the Wireless network over a variety of carriers and data cards (modems). Seems to work well for T-Mobile in particular.
http://www.ibm.com/developerworks/library/wi-enable.html —Includes WiFi, Bluetooth, and GPRS connectivity in Linux.

A.8 MISCELLANEOUS LINKS

http://en.wikipedia.org/wiki/WURFL—A description of the Wireless Universal Resource File open source project that provides information on the capabilities of lots of wireless devices (mobile phones).

http://people.csail.mit.edu/rudolph/Teaching/Lectures07/—Professor Larry Rudolph's online material for a pervasive computing course that contains lots of useful ideas which can be used to create automated tests run from mobile devices.

http://www.geocities.com/model_based_testing/—Harry Robinson's web site containing useful information on Model Based Testing. Although the home page says the site was last updated in 2004, there is more recent content, e.g., from one of his presentations in 2006. It is well worth a visit.

http://www.adobe.com/products/flashlite/—FlashLite from Adobe.

http://www.waptutor.org.uk/—An extremely basic introduction to creating a WAP page in WML.

http://www.w3schools.com/wap/wap_intro.asp—A more detailed introduction to WAP and WML, the site includes a useful WML reference.

http://www.openmobilealliance.org/tech/affiliates/wap/wapindex.html—Far too many links on WAP specifications and other technical material.

http://www.w3.org/2006/07/Mobile_Web_Design.pdf—An attractive introduction to the mobile web.

A.9 COMMON TOOLS

http://www.wireshark.org/—Homepage for the free and very powerful Wireshark protocol analyser.

http://www.gosymbian.com/fexplorer_new.php—Homepage for FExplorer for Symbian Series 60 phones.

A.10 OTHER REFERENCES

HTTP Pocket Reference, O'Reilly, Clinton Wong, ISBN 1-56592-862-8

Python phrasebook, Developers Library, Brad Dayley, ISBN 0-672-32910-7

www.python.org—Includes the Python software and extensive documention (a tutorial, library reference, etc).

www.diveintopython.org—A brilliant free online book, which is also available in print. Intended for programmers who have experience of programming in another language: read it anyway and work through the examples.

A.11 RAW INGREDIENTS

If you are wondering what you'll need to automate your tests, here is a list of raw ingredients you might need or use:

- Experience in programming in one or more languages. You may choose to learn the language used by the developers when creating the applications or scripting languages that help you to test independently. Some common choices are:
  - Java,
  - Python,
  - Java ME (originally called J2ME),
  - You may also need various flavours of C++ for some native platforms, e.g., for iPhone, for Windows Mobile, and for Symbian.

- Hardware
  - Computers,
  - GPRS (etc.) modems,
  - Phones (that the software is targetted for),
  - Automation devices that include physical phones,
  - Digital cameras.

- Data, which can come from several sources both externally (e.g., from public and commercial sources) and internally (e.g., from log files).
  - The Wireless Universal Resource File (known as WURFL) that contains details of the capabilities of many devices;
  - The commands used to interact and control modems (known as AT commands as the first two characters of the commands are the letters AT). Some commands are limited to a subset of devices;
  - Network configuration parameters, such as the Access Point Name (APN) used to establish a network connection by a device;
  - Web sites, where individuals and organizations have shared information relevant to mobile devices; and
  - Logs (for example logs recorded by web servers may contain device header information generated by devices with every web request they make).

- External software
  - Run-time emulators, device and platform SDKs,
  - Protocol emulators,
  - Protocol analyzers,
  - Utilities,
  - Browser plug-ins,
  - Unit testing frameworks, and
  - Open source libraries (e.g., for Python).

APPENDIX B

Data Connectivity

Virtually all mobile phones include a data modem used for data communications. These modems support a standard form of commands based on the venerable Hayes AT command set from the 1980s, e.g., ATD 12345 would ask a modem to dial the phone number 12345.

GPRS modems and the more modern 3G modems support a common set of extensions to the basic, older AT command set. The extended commands have a + after the initial AT. For example:

```
AT+CGDCONT=1,"IP","Internet"
ATD*99***1#
```

The AT+CGDCONT is an extended command used to configure the modem with the carrier's network settings called the APN, while the ATD is a standard Hayes command to dial the special number used by this carrier (and many carriers) to connect to the GPRS network. The 1 towards the end of the number tells the modem to use the first set of APN values, which was configured by the AT+CGDCONT.

Note: your GPRS modem or phone may support additional commands using other special characters, e.g., AT#

Try to obtain a copy of the AT command reference for your phone or modem; the ones from Siemens are particularly good.

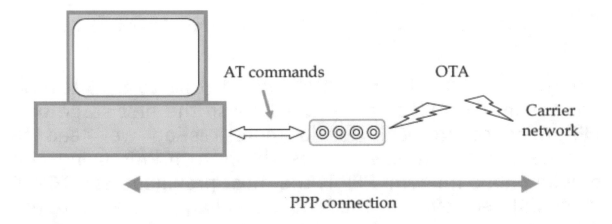

AT commands OTA Carrier network

PPP connection

DIAGRAM: AT commands are sent to the modem to establish a connection. PPP is used by the computer once the connection has been established to provide network connectivity.

B.1 HOW TO USE A MOBILE PHONE MODEM INTERACTIVELY

The modem commands are not available from the UI of the phone, instead the phone needs to be connected to a computer (or a similar device). The connection can be hardwired (e.g., using a USB or serial port cable), wireless (e.g., using Bluetooth), or over an infrared connection— depending on which features the phone offers. The connection is represented as a serial port in common operating systems, including on Microsoft Windows and Linux. The operating system may allow you to configure the identifier of the serial port (e.g., the COM port number in Microsoft Windows). Details of how to configure the serial port and how to get started with a terminal emulator are available in Appendix C.

Once the serial port has been correctly configured and tested, you should be able to use a terminal emulator program to enter commands directly.

B.2 HOW TO USE A MOBILE PHONE MODEM FOR IP TRAFFIC

Using AT commands limits the type of test to SMS testing and testing an APN's ability to establish the first stage of a PPP network protocol. To do more testing we need to configure the computer to establish a TCP/IP using the mobile phone modem. PPP is the most prevalent way TCP/IP is configured over a modem, including mobile wireless devices. Please see Appendix C for details on how to configure PPP for Linux and Microsoft Windows.

B.3 POSSIBLE PROBLEMS WITH DATA CONNECTIVITY

- Phone configurations (APN settings);
- Using limited APN configuration, e.g., one that does not allow UDP, using WAP vs. "Internet";
- Poor network conditions, e.g., signal strength, network congestion, GPRS vs. 3G, etc.;
- Blocking of:
    - IP protocols, e.g., UDP on some price plans or carriers, and
    - Some application protocols being blocked by operators, for example until the summer of 2007 RTSP was not available on at least one UK network operator;
- Inappropriately transcoded content; and
- Walled gardens (where the carrier restricts users to an often very limited subset of the internet).

B.4 MISCELLANEOUS PROBLEMS

- Inappropriate caching by operator networks,

- Cookie issues (ranging from phones not supporting them at all, to limits on the number and/or size).

Configuring Your Machine

These are the main steps to configuring your machine to use a mobile wireless device such as a GPRS modem or a phone:

1. Prepare the device,
2. Connect the modem or phone,
3. Establish communications using a terminal emulator program, and
4. Configure PPP so your computer can use the connection for network (TCP/IP) communications.

This appendix provides brief instructions on how to get started.

C.1 RAW INGREDIENTS

- A computer with either a GPRS data card or a GPRS modem. I have had good results using an external Siemens MC35 GPRS modem, cost is around $250 given the current US exchange rate (£150 GBP).
- The SIM.
- The APN settings.
- Knowledge of either Linux or Windows dial-up networking, modem configuration, etc.
- Patience!

C.2 PREPARE THE DEVICE

- Put the SIM in to the GPRS card or GPRS modem.

- Connect the GPRS device into the computer.

C.3 CONNECTING YOUR MODEM OR PHONE

There are various ways a modem or phone can be connected to your computer. External GPRS modems are generally connected using a serial port and a serial cable. PCMCIA cards integrate into your computer and generally need a software driver in order to operate. In Windows they appear as one or more serial ports, in Linux they appear as one or more /dev/ttyUSB devices.

Phones can be connected using a USB cable, infrared, Bluetooth, and possibly even other ways I have not seen yet. The choice depends on the capabilities of your computer and phone. For instance, both need Bluetooth in order to use it. I recommend using the USB cable if it is available. In Windows a software driver is generally required.

Serial ports need to have their speed and other properties set correctly on the computer in order for the modem to respond. Try setting the parameters to 115200 bits per second (also incorrectly called baud), 8 data bits, no parity, one stop bit, and hardware handshaking (also called RTS/CTS handshaking).

C.4 USING HYPERTERMINAL IN WINDOWS

HyperTerminal is available is Windows 98 and later versions of the operating system. It provides basic terminal capabilities and can be used to interact with your phone or modem.

http://www.developershome.com/sms/howToUseHyperTerminal.asp—How to use the HyperTerminal in Microsoft Windows to control a modem, e.g., to enter AT commands for a GPRS modem.

C.5 USING MINICOM IN LINUX

Minicom is one of the many possible terminal programs in Linux. The manual should be available online (type man minicom to load the manual). A combination of control keys gives you access to the menus. Control+A followed by Z displays the main menu. O loads the configuration menu, then select the Serial Port setup where you can select the device.

/dev/ttyS0 represents the first serial port, and /dev/ttyUSB0 tends to work for a PCMCIA GPRS/3G data card. "Enter" tends to close menu options. I recommend you save your changes as dfl (I think this means the default to use), then quit and restart minicom in order to use the changes reliably.

C.6 CONFIGURING PPP IN WINDOWS

I have removed from the current document to save space. Details are available from the author.

C.7 CONFIGURING PPP IN LINUX

In Ubuntu Linux (and I expect most other flavours) all we need to do is create a configuration file for wvdial and then call wvdial with the relevant configuration name. I am assuming you will want to create separate configuration settings for each carrier you may want to test. Here is an example with two Vodafone UK sections, one for contract SIMs and the other for pay-as-you-go:

```
root@2atestuk01:/etc# cat wvdial.conf
[Dialer Defaults]
New PPPD = yes
Modem = /dev/ttyS0
Baud = 57600
Init = ATZ

[Dialer VodafoneUKContract]
Phone = *99***1#
Username = web
Password = web
Init2 = AT+cgdcont=1,ip,internet

[Dialer VodafoneUKPAYG]
Phone = *99***1#
Username = wap
Password = wap
Init2 = AT+cgdcont=1,ip,pp.vodafone.co.uk
```

Obtaining the Username, Password and the specific AT command used in the Init2 strings are often hard to find. I have frequently spent hours searching on the Internet for the relevant details for specific combinations of network operator and data plan.

To start the connection, call wvdial followed by the name of the section you want to use, e.g.

```
wvdial VodafoneUKPAYG
```

If not, you can simplify your configuration by only having the:[Dialer Defaults] header with one set of Phone/Username/Password/Init2 values. For example:

```
root@2atestuk01:/etc# cat wvdial.conf
[Dialer Defaults]
New PPPD = yes
Modem = /dev/ttyS0
Baud = 57600
Init = ATZ

Phone = *99***1#
Username = web
Password = web
Init2 = AT+cgdcont=1,ip,internet
```

To start the connection, call wvdial as follows:

```
wvdial
```

Here is a typical log of a successful connection:

```
--> WvDial: Internet dialer version 1.55
--> Initializing modem.
--> Sending: ATZ
ATZ
OK
--> Sending: AT+cgdcont=1,ip,pp.vodafone.co.uk
AT+cgdcont=1,ip,pp.vodafone.co.uk
OK
--> Modem initialized.
--> Sending: ATDT*99***1#
--> Waiting for carrier.
ATDT*99***1#
CONNECT
~[7f]}#@!}!}#} }9}"}&} }*} } }'}"}(}"}%}&R!}4O}#}%B#}%n[02]~
--> Carrier detected.  Waiting for prompt.
~[7f]}#@!}!}#} }9}"}&} }*} } }'}"}(}"}%}&R!}4O}#}%B#}%n[02]~
--> PPP negotiation detected.
--> Starting pppd at Mon Aug 20 15:40:27 2007
--> Pid of pppd: 8271
--> Using interface ppp0
--> pppd: È¿EÀI[06][08]`H[06][08]
--> pppd: È¿EÀI[06][08]`H[06][08]
--> pppd: È¿EÀI[06][08]`H[06][08]
--> pppd: È¿EÀI[06][08]`H[06][08]
--> pppd: È¿EÀI[06][08]`H[06][08]
--> pppd: È¿EÀI[06][08]`H[06][08]
--> local  IP address 10.172.165.66
--> pppd: È¿EÀI[06][08]`H[06][08]
--> remote IP address 192.168.254.254
--> pppd: È¿EÀI[06][08]`H[06][08]
--> primary   DNS address 10.205.65.68
--> pppd: È¿EÀI[06][08]`H[06][08]
--> secondary DNS address 10.205.65.68
--> pppd: È¿EÀI[06][08]`H[06][08]
```

To test the connection use the ping command specifying the newly created PPP network, e.g.,

```
ping www.google.com -I ppp0
```

Note: Your current LAN connectivity may play up while you have an active PPP connection. Fixing the routing so both LAN and PPP work concurrently would be desirable; however it is currently outside the scope of this material.

To disconnect press Control-C in the terminal window and the program should respond along the following lines...

```
Caught signal 2:  Attempting to exit gracefully...
--> Terminating on signal 15
--> pppd: È¿EÀI[06][08]`H[06][08]
--> Connect time 3.4 minutes.
--> pppd: È¿EÀI[06][08]`H[06][08]
--> pppd: È¿EÀI[06][08]`H[06][08]
--> pppd: È¿EÀI[06][08]`H[06][08]
--> Disconnecting at Mon Aug 20 15:43:51 2007
```

That is it, good luck!

Author Biography

Julian Harty is currently a senior test engineer at Google, UK. Over the past 3 years his main focus has been testing mobile phone applications, particularly in the area of test automation.

He has presented a number of tutorials in this area in Europe, Canada, Australia and New Zealand.

Julian is also actively involved with the British Computer Society's Specialist Group in Software Testing (BCS SIGIST) in the important area of non-functional testing. Prior to joining Google he ran his own successful software testing consultancy company (Commercetest) in the UK.

A frequent author and speaker at software testing conferences and workshops on specialist subjects including: mobile, productivity, non-functional, performance and security testing, Julian is an internationally respected and sought-after software tester.

Printed in the United States
by Baker & Taylor Publisher Services